A Blade of Grass

Mike Hodgson

Grosvenor House
Publishing Limited

All rights reserved
Copyright © Mike Hodgson, 2020

The right of Mike Hodgson to be identified as the author of this
work has been asserted in accordance with Section 78
of the Copyright, Designs and Patents Act 1988

The book cover is copyright to Mike Hodgson

This book is published by
Grosvenor House Publishing Ltd
Link House
140 The Broadway, Tolworth, Surrey, KT6 7HT.
www.grosvenorhousepublishing.co.uk

This book is sold subject to the conditions that it shall not, by way of
trade or otherwise, be lent, resold, hired out or otherwise circulated
without the author's or publisher's prior consent in any form of binding or
cover other than that in which it is published and
without a similar condition including this condition being imposed
on the subsequent purchaser.

A CIP record for this book
is available from the British Library

ISBN 978-1-83975-102-8

mandjhodgson9@gmail.com
www.abladeofgrass.co.uk

Why is this book called **A BLADE OF GRASS!**
To find out why? You will have to read this book.

Foreword

This thought-provoking book addresses some of the most fundamental questions of human existence and our relationship with God and the world around us from the author's own layman perspective. It takes on Darwinian Evolutionist theories, and emphasises the importance of a missing link between fossils and humans and other incredibly complex living organisms we find in the world today. Mike Hodgson also challenges us to look differently at the structure of even the most simple of things from a blade of grass to a single human cell and unearths the beauty, intricacy and simplicity in them which provides further evidence of an inspired Creator. This reflective book also looks at other scientific tenets wrapped up in the Big Bang theory and Relativity Theory by looking differently at our shared histories and the relationship between space and time, asking if they are human constructs or actually an integral part of the fabric of the universe. He also touches on the spiritual and supernatural aspects in the world and tries to make some sense of how they are all inter-related.

This book holds the keys for future explorations from the layperson to the scholar interested in knowing more about the nature of the world around them and their roles within it, providing them with insights into the creation and meaning of life. If this book compels you to cast aside some of your previous assumptions about science, physics, history, philosophy and religion whilst inspiring you to ask more questions, which I'm sure it will do, then it has truly served its purpose.

Darin Jewell

Ph.D. in Philosophy of Religion, University of Cambridge
Senior Fellow, Center for the Study of World Religions,
Harvard University

INTRODUCTION TO OUR AMAZING WORLD OF WONDER

We are fearfully and wonderfully made. Who can fathom the depth of our world? Did it all come out of the Big Bang? I don't think so. If you were the maker, you would have to know how our brain works and how to create a simple blade of grass. Is there anything missing in the world's creation? Would you have added or taken away anything to make it a more beautiful and better world?

In this book, I have endeavoured to search out the intriguing mysteries that govern the meaning of life, including what I believe you may call new theories or even new revelations that could cause us to rethink and change the way we view our world.

A Challenge to the Evolution Theory

Two main areas I wish to dwell on: firstly, where the first living cell or cells came from; and how many people consider that those cells have evolved and mutated until, ultimately, they developed into you and me, meaning that we are related to every other creature – ant, bird, fish, lion, or dinosaur, and so on.

For more than 150 years, scientists have said that apes are our closest relatives, and although the missing link has not been found, some say we are close to finding it. Does it exist? If we look back to all living creatures and geographical findings, every animal has its own unique DNA. As far as I know, no-one has found conclusive evidence of mutation that has linked one animal to another. Is there an animal, which is part hare and part rabbit, for instance? Or an insect which is part wasp and part bee?

We have given every animal, both alive and extinct, its own identity by naming it. Every time we find a new species, it has its own DNA and is unique. There is not one that we can say has mutated or is in the process of mutating to another.

And if you factor in that many scientists believe that 99% of all living creatures have become extinct, and we are the product of under 1%, then evolution with mutation at some time must have multiplied at an alarming rate to give us what we have now.

If we go back to the first living cell or cells: why did these cells decide they were going to be part of the animal kingdom or part of the fauna and flora? Were there two different types of cells?

I find it difficult to believe these cells came into existence by pure chance. These first cells must have had the intelligence of a thousand computers built into their DNA, which only leads me to the one conclusion: that there must be an intelligence behind our existence. If you also add into the equation that out of all the galaxies and billions of stars, earth is as far as we know the only planet able to sustain life. Are we one-in-a-billion, or even unique?

Was our world and planet created by a creator, or could it be a mixture of both, i.e. did a creator make our world so it had the ability to evolve and give us nature, therefore it had Natural Selection built into its DNA? Or is there another reason for our existence?

I will also endeavour to explore whether Charles Darwin's theory in his book *Origin of the Species* stands up to scrutiny. Or does modern science tell us a totally different story, taking into consideration that we now have the benefit of DNA that can identify each and every living creature?

Also, in this book I will explore the 'supernatural/natural' and discuss whether they are intricately woven together? I will endeavour to try and unravel some of these questions.

Why was 6,000 years ago the start of a new beginning in the Earth's order?

This is another of the main reasons for writing this book, and its importance in the world's history. If we look back just 250-300 years ago, we did not have access to the depth of knowledge of science, history, and geology of what existed pre-6,000 years ago.

There are many events that emanated from the period 4,000BC, but they all centre around the fact that man was given an intelligence that far surpassed any prehistoric animal or man. Before that time, man lived mainly in caves, had no concept of right and wrong, and lived with the nature of the survival of the fittest. Geologists have calculated that the first prehistoric man came into existence some 225 thousand years ago.

So, why this sudden change? Everything fundamentally changed with the coming of this modern man, who had a cognitive and supernatural element to his nature, compared to his ancestor? He had an audible language, the ability to reason, artistic skills, a creative ability to build houses. And in the space of just 1,500 years, he was building lavish palaces along with exhibiting many other talents.

The Bible is considered to be our oldest historical book, and it just happened to take the records of the Hebrew Nation's history to the same era 6,000 years ago. In the first book of Genesis, it says that God created man in His own image and breathed His Spirit in him. If we have a creative God, then it stands to reason that this new man has this creative element built into his being.

In this book, I will look at other events that have happened throughout our modern-day history. But, ironically, they all fall within the 6,000-year perimeter – every civilisation, every religion. Yet nothing before.

Other areas covered in the book.

What is Time and Space?
Without these two elements there are no building blocks for our existence, however, there is no known evidence to prove a beginning and end of both of these.

There is another theory, that they always existed with no beginning or end. Or could there be another dimension outside our timescale that ordered what we are?

The Big Bang Theory
It is said this Universe, as we know It, started with a big bang that flung the planets and galaxy into their orbits.

This, our beautiful and amazing world
The book will take a look at the many wonderful elements that make up our Universe, and some of its facts and mysteries.

The Bible
Is the Bible the true recorded history of the Hebrew Nation, and is it our earliest history book? Was it inspired by God? What evidence is there? What do the Dead Sea Scrolls tell us, and do they give the Bible credence?

Who was Jesus?
Was Jesus historically real, or just a myth? Was He the Son of God, and what was the evidence? And how His three-year ministry, 2000 years ago, set the platform for our future world.

Spirit and Soul, and a hope for life after our physical death

Science cannot quantify them. Where did they come from? What are they?

It is generally believed we all have a Spirit and Soul; it is that supernatural God-part that defines who we are.

Man changed 6,000 years ago. His Spirit took on a new dimension. God gave this new man a new part to his Spirit – the ability to choose to accept God as his Creator. Man was provided with the understanding and knowledge to either reject or accept Him. And if we accept Him, we will take on an eternal part to our nature.

Believing in the natural/supernatural

Is there a pure world based on science and physics where every question can be answered? Or is there what we call a supernatural force that created everything we know? Or could it be a combination of both? Maybe if there is a supernatural force and if we understood more about it, it would no longer be known to us as supernatural, but as the norm. 'Natural'.

That also brings another thought into question. If everything was and is created by a supernatural being, then does it follow that even a blade of grass and grain of sand is supernatural – or the norm?

Is there life in another dimension?

Is there a life outside our world? Heaven? Is there a hope for a future life? If not, everything is meaningless, Why was life created in the first place? Is there a God who created us?

My final thought for you to try to comprehend is…

Does time and history exist? **IF** – and that is a big 'if' – consider this:

Because we are all just a part in passing time, and presuming this world's civilisation that came into existence millions of years ago will one day come to an end – either by a natural catastrophe, by our own doing, or by Armageddon – then it stands to reason that time and history will not exist! History is only history if we have the cognitive ability to understand what history is! To all animals and prehistoric man, history is meaningless. Did God give us history, or is it an invention of man?

This book sets out my path over the last six years in discovering what is the reason for life… beyond just being a Christian.

In the last few years, I've been questioning what evidence there is for Evolution. Or if God created this beautiful world of ours. As I continued to explore, I found more and more evidence for a supernatural world, but this only provided me with more questions. Talking to people, I found that some of them were very clear in what they believed, but the majority did not know what they believed. That included agnostics, atheists, Christians, and people of other religions. I needed answers!

There are many intellectual scientific and philosophical books written about the meaning of life, which are far beyond my understanding. Some are fact, some fiction, or a mixture of both.

Although I'm a committed Christian, I have endeavoured to explore the evidence objectively, and this has driven me to explore some of the claims that have been made in the past. The first thing I explored involved the Bible, which is 'considered to be our oldest history book' and what credibility it had. This led me to look closer at what we know about the origins of the world, which provided me with even more questions.

I'm not a scientist or theologian, but what I have written is from an informed layman's position. I have based it on what to me makes logical sense. **Logic is the balance of probabilities!** For example, in the approach to making decisions about the big questions, you could make the assumption and argue that it is rational to accept a more likely of two or more explanations, in the absence of any other evidence. However, I do believe I have been inspired in many ways – be it through divine revelation or not – to write and include some of the book's contents, but you can make up your own mind.

I hope this book will speak to people from all walks of life, including academics. It includes basic science, physics, geology, history, philosophy, and religion, but is not too difficult to understand.

I believe it will give people evidence and conjecture to enable them to have enough information to come to a logical understanding and belief in what is the true meaning and reason for life. And even if you don't come to a decision, at least I hope your life will be enriched by its contents.

In the last ten years, I've watched many documentaries which have been made about our wonderful world, exploring history and natural history, science, geology, physiology, theology, and so on. Some are based on what we know and can prove; others are based on conjecture, assumption, and theory, in the fundamental human search for the truth. There are many factors about our world that we are probably never going to know. I realize what this book is trying to achieve will be very controversial, exciting, and some may say brave, but we are all on a journey with a natural desire to find what life is all about.

I hope this book will be easily understood by the majority if not all readers.

Contents

1)	In the Beginning theory	1
2)	Creation or Evolution/Natural Selection	10
3)	Prehistoric and Modern Man	28
	Prologue to the next chapters of the book	32
4)	The Birth of Human Civilisation and the Uniqueness of Modern Man in the Hebrews	40
5)	Who were Adam and Eve?	48
6)	Is the Bible credible?	55
7)	Who is Jesus and how has Christianity changed our world?	80
8)	Is there a force for good and evil (bad)?	104
9)	Mind, body, and spirit?	106
10)	Believing in the natural/supernatural	119
11)	Life in another dimension	121
12)	Last thoughts: What is History?	123

Chapter One

In the Beginning theory

Before you start reading this book, it is aimed to stimulate your mind and take it to places that you may never have dared venture or ever thought of delving into, including areas that will touch on the secrets and the meaning of life.

There are many theories of how the world began. Did it start with a big bang? Well, it all depends on what you define as the world. Was it an empty void? Was it filled with matter? Did it have any form of life, even if it was just a tiny cell?

Let us look at the Big Bang theory! First of all, there must have been time! Then there must have been space! And there must have been matter. Without these three elements, we have no starting points. However, was this just a starting point in the world, or does it go back much further? We must start somewhere, and have to accept we will probably never know how these three elements ever began or were created, as our minds cannot comprehend infinity. So, does this mean we cannot understand God?

The Big Bang? First scientists say it happened some 14 billion years ago? What was it? Well, time is established! Space? How do you quantify 'space'? You can say it was X metres long or so many light years away. Was earth at the centre of the Big Bang, or were we just flung out to the edge of the Universe, or are we in the very centre? We haven't even established what the Universe is? Is it quantifiable?

Recent scientific discoveries have found that the Universe is expanding at an alarming rate. If we take it that the Universe is expanding – there is the theory that it must

– then, 'some 14 billion years' ago it would be decreasing and decreasing until the Universe was just a minute speck. That speck must have contained incredible energy, heat, and the ability not only to explode and expand but also had the incredible intelligence written in its DNA to create the world we know. Also, space would have to increase to house the transforming Universe! Our minds cannot just comprehend infinity, or the infinitesimal; it's like circumnavigating the earth forever… whatever forever means.

To have a big bang, we must have matter, and the earth is made up of matter! Was there a huge planet made up of all the stars we know of in our solar system, which for some reason exploded? Or was it so compressed into a small speck that when it exploded it expanded? What made it explode? What did it explode into? Space?

It's strange how these huge – to us – lumps of matter circumnavigate one another; their course does not change! Scientists have calculated that if the earth, the moon, or any other planet in our solar system was one mile adrift, the whole system could collapse. They are held in place by gravity! And other forces? We know the effects of gravity, but scientists have not discovered how exactly how it works. They have now discovered what they call 'Dark Matter' and 'Dark Energy', but they have not found a way of proving what they are or how they work, and what place and use they have in our Universe. Could they play a part in holding the Universe together?

If the world before the Big Bang was just a lump of rock in the sky, so to speak, with no particular order, then how did amazing order come out of a big explosion? Explosions cause disorder and destruction! How is it that all this just happened? Was there some other event that caused this world to suddenly appear in the sky? We do not know, and probably never will. Or maybe there has to be a Creator/Designer? I cannot think of any other explanation.

Let us now get back down to earth, so to speak. If we look at this amazing planet we live on, compared to all the other planets in the Solar System, we can see that the earth is special in many ways. It is made up of many components which are needed to sustain life. First of all, we have this amazing substance called water. If we were a product of the Big Bang theory, one would have thought that there was water or vapour that must have existed somewhere in the explosion. Does it continue to exist on other planets? If it does, where does it come from?

Scientists have explored other planets and believe there are many planets with water on them. But if so, have they the ability to sustain life? The answer to that is they would have to possess an eco-system very similar to ours. And when you think of all the elements that are required, it is immense: the exact gravitational pull; the right oxygen; the correct balance of water and climate. And surely that must involve an intelligent force – a planner/creator – to make it work? Could it really be just by chance?

One other aspect is that on other planets in the solar system there is a distinct lack of the colour green. Apart from minerals, this points to the non-existence of any form of life. Astronomers have discovered a planet some 120,000 light years away that might have the correct eco-system to support life, but it would take us more than 120 years to reach, using our fastest space ship.

Green is the dominant colour to our planet earth, but on other planets the colour green is thought to be from mineral structures. It is mainly created from plant life, and without water and the right eco-system the leaves will soon turn brown, wither, and die. When the first American astronauts went on their journey to the moon, they looked back at earth and exclaimed how beautiful our planet is. If you look at the sea, it either looks blue reflecting the sky above, or green reflecting the life below the surface. This is truly a beautiful world we live in.

Water

If we again look at water, how would you transport billions of litres of water from one place to another? Ah! I'll invent clouds. But to do this, there has to be wind – 'not too much and not too little' – to blow the clouds. OK, then there has to be gravity. And that would have to have exactly the correct force to allow the water to fall at the right speed. Too heavy, and it would crush plant life and damage what has been created; if it was too light, the rain would remain in the air and be like mist in the clouds that hold the water, and would possibly just float as on the surface of the earth.

On the moon, the gravity is far less than on earth, and perhaps that is why there is no water there. Perhaps it has all evaporated. Or it's held in ice. However, another problem with water is that if it evaporates, it still exists and must go somewhere. Water, like wine in a bottle, does not evaporate. Is the water trapped on the earth by the atmosphere?

One of the other features of water is rainbows. OK, we know they are produced by a reflection from the sun's rays, but why is the spectrum of colours in a specific order? I'm sure there is a scientific reason. They are truly so amazing and beautiful, and are also known to have appeared on special occasions. By pure chance?

Gravity

Let's look at gravity. If gravity was greater, then all living creatures would have to change and adapt to a totally new structure. Our blood pressure would be a lot higher, so all the working parts in our body would have to be a lot stronger to allow us to move freely, and we would have to exert a lot more energy to even walk. Birds and aircraft would have to be a lot stronger to fly, and we would need a lot more petrol in our cars and aircraft to get enough speed out of them to defy gravity. It would be almost impossible to swim and stay afloat. And diving would be too dangerous. If you hit the water at a speed of 20% faster, you could be seriously injured and would not have the time to tumble or twist. And so the list continues.

However, if our gravity was less – like it is on the moon – we would have a totally new set of problems. The effort to walk would be far less exerting but it would take so much longer to get anywhere. The laws of force against resistance would come into issue, so athletes' Olympic records would be different! They would have to be recalculated. And I'm sure scientists would tell you of many other problems we would have to deal with.

There is another phenomenon about gravity that scientists with all their knowledge have still not discovered: exactly how does it work? A magnet can pull metals towards it, but gravity pulls everything on earth towards the centre. Is it a pull from the core of the earth? Or could it be a force that comes from above? Or a mixture of both? If you hold your hand out in front of you, is it being pulled down? Or being forced down from above?

We know the gravitational pull from the moon affects our tides. How is this planet of ours being kept in its right orbit? If our gravity, or the moon's gravity, was greater, surely we would be slowly pulled together and we would collide with one another? That is why I dare to suggest that there are other forces in place that keep us in perfect orbit around the sun and moon.

If the moon did not exist, we would not have tides, and our oceans would be stagnant pools of water. Our waters would not be oxygenated to create plankton to recreate oxygen, and all life would cease! Take a minute to consider this: the sun is 400 times larger than the moon; the earth is only roughly four times bigger than the moon; and 1.3 million earths would fit into the sun. It is quite amazing that our solar system is so arranged that we have solar or lunar eclipses that are, on occasions, perfect in shadowing with the sun! They also give us day and night, and are the correct distance to give us the perfect temperature to sustain life. If the earth was 10% larger or smaller, we would not exist! Some of the latest thoughts and discoveries are that gravity is created by 'Space Time', whatever that is.

Light and heat

These are two other elements that have to be right for us to exist! That is why the sun is the exact distance from the earth: first, to give us the mean temperature not to burn us and nature to a cinder; and also to give perfect light for us to see clearly. The ozone layer, some 40 miles above the earth, protects us from harmful rays and stops us burning up. Without it, we would not exist, as it allows the correct heat to filter through. It's just another part of this world's amazing make-up.

We know that living creatures do adapt to the environment in which they live, even humans. For instance, the nearer the equator, the darker our skins will be to protect them from the sun's rays. Although we have adapted our ways to live in extreme weather conditions, there is a limit to what our bodies can tolerate. We have now worked out that our climate is perfectly balanced, and if it changes by just a few degrees it will have catastrophic effects on our world.

We know the earth has a way of keeping itself in balance, but are we pushing the balance so far that we are in a danger of destroying the earth with climate change? We know we have developed the power to destroy the planet with the atom bomb, but what does that mean? If earth was blown out of existence, would that upset the whole order of the Universe? Or if its surface was only destroyed, would the order of Universe still remain? Did we come from chaos or design?

My question to the above is: how did the earth get to such a condition or evolve so that it could sustain all living creatures? Or was it made especially for humans? Is it too incredulous to believe that all the elements of life had to be perfect before we could come into being and to survive?

Other forces

With all our intelligence, we still have a lot to learn about the basic structures of earth. It is quite incredible, and we have to acknowledge we live on an amazing planet in an amazing world. Professor Brian Cox stated that some 4.6 billion years ago a meteorite collided with the earth and knocked it to an axis of 23 degrees. That is quite amazing that it was pushed to the exact degree that would enable the whole of the earth to be under the gaze of the sun at some time in the seasonal year. One degree above, and it would have been too much; one below, and it would be too little. That's assuming we did collide with a meteorite in the first place.

I would dare to suggest that we are tempted to find a formula that fits. However, if earth was not set at that angle, we would not have seasons; some areas of our planet would be in constant darkness, and others in constant light; in some areas there would be frozen wasteland, others would be too dry to maintain life. I'm sure earth would be an entirely different place to live in, if it could actually sustain life.! If any of the elements such as water, clouds, light, heat, wind, and gravity, were either not present or were at a different level of force or volume, this world of ours would not be able to sustain itself, and life as we know it would just not exist.

I've tried to avoid suggesting that the world was created by a supernatural intelligent force, but whichever way we turn, this world of ours is simply incredible. Did all this just happen by chance? How does something come from nothing? And we haven't even touched on the creation of life! It is like taking all the words in the dictionary or Bible, dropping them from a height, and them all falling down in the correct order. Even the word 'creation' says it is made, 'created', so must come from some source of order. If something is made, there must be a creator – a force far beyond our comprehension!

Take some of the most formidable scientists and philosophers of our time. I've listed just a few of them here: Francis Collins was pictured on the front cover of *Time Magazine* for his work heading The Human Genome Project, and wrote the book *The Science of God* – he explains how science led him to believe in a creator; Sir John Polkinghorne is Cambridge's top quantum physicist, a member of the Royal Society and an ordained Anglican priest – he speaks of his belief of a creator being compatible with science; and finally, Dr William D Philips, who won the Nobel Prize for physics in 1997, has spoken widely about how his dedication to science and God are not merely compatible but conjoined and logically inextricable from one another.

Many other people believe that science and a belief in a creator are not at odds with one another, but it is hard to justify that this world just created itself out of thin air, so to speak, and that it just happened with no plan or design.

So, taking in all of the above, I am going to suggest that if even some of the aspects of the Big Bang Theory may be true, there had to be a designer and planner in charge of the world. It's just too miraculous to be down to chance. Many scientists believed in a creator and that science only points to a supernatural creator. Such famous names as Michael Faraday, Isaac Newton, John Clerk Maxwell, and Galileo. Even Albert Einstein wrote, *Science can only be credited by those who are thoroughly imbued with the aspiration towards truth and understanding. This source of feeling, however, springs from religion. To this there also belongs the faith in the possibility that the regulations valid for the world of existence are rational, that is comprehensible to reason. I cannot imagine scientists without that profound faith.* He also said, *Science without religion is lame; religion without science is blind.*

I'm sure there are many other famous scientists who would argue against the existence of an intelligent being that created our world. This age-old debate will go on and on, long after we depart this world. It is up to each of us to form our own ideas, based on whatever is known, be it science, physics, historic findings, or what is yet to be discovered. But whichever way you turn, there will always be an element of faith!

It takes a great leap in faith to believe in a God who created everything, a God who cannot be physically seen, a God who allows suffering. But I would suggest it would take more faith to believe it all happened by chance, without any outside influence.

Later in the book, I will talk of the rationality in believing in a supernatural being (God).

Chapter Two

Creation or Evolution

The words Creation, Creator, and Creature are intricately linked together, and all exist with one another. This also suggests that both Creation and Creatures had to be Created. Before Charles Darwin's book *Origin of the Species,* the above analogy would not have been in question, as the theory that we all evolved from an inanimate source had not previously been considered. However, there were several other scientists of that time who were exploring Evolution theories.

Neutrons and proteins came together at a perfect time when the atmosphere was right for the first living cell to survive and develop? I have a big problem with this question. Why? Who? How?

Why should anything happen at all? Had anything happened previously to set precedence for it to happen again? There must be some reason, or is there no reason at all for our existence? When I look at the world and see the beauty that surrounds us, is that not a reason in itself? Can you not imagine a world without love? Is that not again a good reason? Are we and this world with all its creation, just a chance happening?

I don't believe this. I personally believe everything has a reason and a purpose for its existence. Or are we just a part of pre-selection and, like robots or computers, pre-programmed to do what we are told? Or, unlike robots, do we have freedom of choice to go, think, and do what we want?

I choose to believe we are part of a creation that was brought into existence for a reason, and one of those reasons was man. Imagine if we could get some of the greatest scientists, such as Hawking, Brian Cox, Isaac Newton, and many others, all together, regardless of what they believe. With all their intelligence, could they create a single blade of grass, never mind getting it to reproduce? To say this all just arrived and it all started after the Big Explosion theory is quite incredible!

Who or What created our world? Looking at the last paragraph, it leads us to believe that our world either just happened by chance, or different neutrons came together at the right time. Could there be an intelligent force behind it all? Or maybe something else? Your choice to believe!

How did life start? Well, first the climate and conditions had to be right. It had to be when everything had been prepared at the right time and when all the stars were in the right place, so to speak. The Universe had to be set in motion, whether it be from a big bang or some intelligence source. Something had to set it in place, and then there must have been something that sparked the first living cell.

Was it an already formed creature? Or was it just a single cell? if it was an already formed creature, the word creature suggests 'to create' so there must have been a creator. And, would there have been one form of animal, large or small, or was there a whole range of animals and insects? That seems an amazing thing to believe in, so should we just rule it out of the equation? I don't think so, because the only other alternative is to believe we evolved from a single cell or cells. And for several reasons, I find this just, if not more, incredible to believe.

How does something come out of nothing? First, that cell or cells had to come from somewhere. Could it come from some primeval soup? David Attenborough's BBC programme *First Life* talks about evidence of life that was present in ice during one of the first ice ages, and life was nearly annihilated before it began. These organisms survived, and when the atmospheric conditions were right and the earth began to warm up, these organisms took off and exploded into life. As volcanoes eruptions flew into action, this created a greenhouse effect, heating the water and melting the ice. The water and nutrients flooded into the oceans and grew onto the rocks, which pumped out oxygen and gave rise to the animal kingdom. Microscopic life grew into something bigger. After billions of years, a life cell came into being, and then, David Attenborough said, 'Something amazing happened in the deep sea, with the increased oxygen, causing the cells to stick together, these clumps ultimately evolved into animals.'

It is interesting that David Attenborough stated 'something amazing happened…' and referred to cells sticking together and 'these clumps ultimately evolved into animals'. This is basically saying that these cells somehow developed heads, tails, brains, and reproductive organs. Then, taking events from there, these creatures evolved into all living animals, including humans.

Let us go back to where the first living cell came from. Somehow, this cell or cells came from where? This cell would have to be more intricate and intelligent than thousands of computers put together; have the power to create the human brain with all its intelligence; the ability to implant feelings, love, emotions; to create flowers and all nature; and so the list goes on and on and on. And all in the size of a pinhead cell. That to me is really too incredible for words! However, something amazing had to happen to kick start our world.

There are some people who say we were brought here by aliens, but did they bring the animals as well? This brings us back to the old question: where did they come from? Were they a product of the Big Bang? And another thought: is there life outside our solar system? I personally believe that our planet, earth, is unique and a one-off. But there is no reason to say whether life as we know it, or something similar, could not exist somewhere else in the Universe, whether through a creator or by a scientific reproduction, or some other way we have not even thought of or discovered yet.

We have discovered so much through science, physics, geology, biology, astrology, and much more, but with every new thing we discover, there seem to be even more and more questions.

Let's go back to the theory that we came from an original cell, and follow it through to Darwin's theory in the *Origin of Species*. That tells us that we all evolved from the same family of chimpanzees and that they are our nearest relatives. But look at the difference between them: monkeys can ape man, but they have not got the ability to create things like paintings and music, or to invent cars, aeroplanes, computers, and so on. Unlike modern man, they still live by the basic instinct to survive, in much the same way as prehistoric man. Science has also proved that different species cannot mutate – a cat is a cat, and a dog is a dog.

DNA. We have now discovered through the DNA of life that we are separate species with our own unique blueprint that cannot naturally be changed. We are like a pack of cards with 52 different components, and we need every card to create the human body – 50% from men, and 50% from women. And if any one of these components is missing, the human embryo would not fertilise. Recent experiments have proved this.

We also know that different species of animals have their own individual pack of cards which cannot be changed, and they cannot interbreed to create a new species. That's not to say that animals and humans do not adapt to the climate and conditions they live in. But this is completely opposed to Darwin's Evolution theory! And, dare I say, utterly destroys his theory. Even Darwin in his later life apologised for his theory, because it had led many people to believe it to be the origins of all life?

Chimpanzees have about 97.5% of their DNA the same as humans. When and how did that 2.5% change from one species to another? Was there a time when it was not 97.5% but 99%? When and how did this mutation happen?

Why have chimpanzees got 24 pairs of chromosomes in their DNA and humans only 23? Again, when and how did this happen? Somehow chimpanzees must have lost one of their pairs of chromosomes to enable them to become human, but again, if you take just one chromosome from an embryo it does not fertilise.

Many scientists are coming to the conclusion that it is unlikely that man evolved from apes or chimpanzees.

Over the centuries, some of our DNA appears to have been damaged. and we have inherited defects which we can pass on to our children or which leave us open to different diseases. But now we have the ability and are learning how to repair the molecular structure of the DNA, with the possibility of eradicating some of these defects. This is a very exciting but scary development in our history of life, and who knows what it will be like in another 10 or 20 years? I dread to think what Hitler would have tried to do if he'd thought he had the ability to create a superhuman race of people.

However, maybe there is a danger that we become too clever for our own good. For instance, if this world as we know it, was created and existed for millions of years, is in perfect balance, but in our so-called intelligence we think we could make it better by changing its order (i.e. by genetically modified crops, pest control, or other ways), I believe we are interfering with a world that was created in perfect balance and intricately woven together. We need all the wildlife and insects in particular to maintain that life, as they are our main propagators of our crops. Scientists and biologists have already discovered that our bees have been quite severely depleted, and they put this down to pest control and genetically modifying crops.

Another discovery scientists have made is that the simple worm is also vital to our ecosystem; without them, our gardens would not be irrigated.

There so many other ingredients needed to sustain life on our amazing planet. If we lacked just a few or even one of them, our world would cease to exist. It is in perfect balance and we need to carefully look after it like a precious newborn baby. In some ways it is very delicate, but in other ways it is extremely powerful, able to sustain, regenerate, and look after itself, as it has done for millions of years.

The argument for creation from a creator, or a world that evolved from a single cell or group of cells from some thick soup or something totally different, is really down to each individual's belief. I understand that if you don't choose the Evolution formula, then you would have to consider a creator 'God', whatever that entails. But simple logic tells me this world is so complicated that it must come from a super-intelligent force, far greater than we could even start to imagine.

Let us look at the complex nature of man and his senses. Our eyes are amazing in their invention. They communicate with our brain to tell our mind what is around us, warning us of any oncoming danger, showing us the beauty that is all around, enabling us to be creative and so much more, and connecting us to our emotions, whether it be love, hate, anger, joy, and so much more. And our emotions are so much a part of our makeup. When we cry or laugh, that releases tension and stress – again, all connected to our brain, which is like a super-computer, also giving us a full spectrum of colours.

Then there is the sense of touch, which helps us in the dark and to create things. If we were blind, being able to feel is imperative, and it would be the main sense to help us. Also, touch is not only to protect but in showing love to another.

The sense of smell, again keeps us out of danger or helps us appreciate the wonderful world around us, including the perfume from flowers.

The sense of taste allows us to enjoy food and differentiate what is good and bad.

Hearing: without it, noise would not exist, and it would be hard to communicate. How would we appreciate music and the sound of birds singing?

If you were to create a robot, would there be anything you would add to the human body? More than it has already? Was it all by chance that we were given everything we could ever need or want for? Really, we are beautifully and wonderfully made.

There are many other parts of Nature. Snowflakes. They are always perfect, all different; each one is beautiful and of perfect structure. How can this be? If we look at any flower or leaf, they are all of perfect structure, beautiful, and each one is different. Then if we look at any species of bee, of which there are hundreds, and take just one and examine it under a microscope, it is absolutely beautiful with all the hairs on its body, all the very fine veins in its wings, and a minute heart to pump blood to every part of its tiny body. And their honeycomb is believed to be the most perfect structure.

Take a snowflake, a flower, or a bee, and ask the best artist you could find and ask them to reproduce it on paper. It could never be as perfect as the real thing. And if you asked the most advanced scientist and engineer to produce a small model, less than an inch long, to be able to fly and achieve all the abilities of a bee, would they know even where to start? How can these come from the Big Bang explosion? If our best artist and scientists can't even perfectly copy or reproduce what is already created, never mind invent something similar, how can they just happen, or by a chance get-together made of minute chemicals?

There must be an inventor, creator, or super-being with miraculous power. First of all, we would have to create an environment for them to survive in, and then create the living beings and the structure to allow them to survive and propagate. What do you believe? Some people say we evolved by what they call Natural Selection. But if this was in fact true, where did the nature in Natural come from? The word nature comes from the word natural or vice versa, so it must have started somewhere.

The Amazing Grass Family

If you take a blade of grass, is that not part of a miraculous family of grasses? Most of our food chain relies on the grass family. First, our meadow grass is the main diet for our cattle and much of our wildlife. Can you imagine having no meat from our cattle and no dairy produce? Then there is our wheat, barley, rape seed, and other seed foods, without which there would be no bread or cakes and other food products, and over a third of the world rely on rice as their stable diet. Can you imagine having no grass lawns? And what about our sports fields? It does not bear thinking what our world be like without the grass family! Is it not an amazing invention? Every type of grass has its own individual DNA!

If you got together some of the greatest inventors and scientists like Prof. Hawking, Brian Cox, Michelangelo, and others, could they create a single blade of grass with its reproductive DNA built in?

In the beginning of life, if the first living cell was the start of our natural wildlife – including man – with its multiple DNAs, then there must be another set of cells to create our green food chain, including grasses, fruit-bearing trees, and root-bearing vegetables.

At this point, I'd like to explain that the above section is not the reason for the title of this book; that comes later.

Evolution & Natural Selection

OK, I can understand the theory of Evolution and Natural Selection, and can accept that all creatures could change to suit their environment and develop to protect, to catch prey, and increase their ability to reproduce themselves and multiply. But if we go back to the beginning: if we all came from a single cell or cells and evolved into all the billions of creatures that ever existed, and all from just a small clump of cells or even if there was a huge amount of cells – did they all have the same DNA? Did they all come from a pea soup and had the ability to mutate, grow heads, tails, feet and hands, eyes and brains? Then there are different creatures – mammals and insects – so did they wake up one day and say, 'I think I would like to be a butterfly, or a lion, or maybe a dinosaur, or how about a man'?

When you think that every creature that ever existed and exists today, had or has its own DNA, that's billions of creatures. That's a fantastic amount of Natural Selection and mutation.

I do have another problem, and that is about plant life! Where did that all come from? Did it wake up one day and say, 'I don't want to be an animal; I want to be a fruit-bearing tree so the animals can feed on me, or a blade of grass, so the cattle can graze on me'? Did all plant life come from the same original cell or cells that developed into all plant life?

If we look back at the geographical history of this wonderful planet of ours, dinosaurs roamed the earth for 150 million years. Then suddenly, some 65 million years ago, there was a catastrophic event that killed off some 50% of all living creatures, including every dinosaur. But what about the smaller animals: were they also killed, or did some find a way of surviving? That's assuming they lived alongside the dinosaurs in the first place. Or did Evolution start all over again? I doubt it. Scientists have calculated it would take many millions of years for animals and insects to mutate and evolve. That's if mutation is at all possible in the first place, and scientists have no proof that this can actually happen anyway.

Let us look back at the chameleon, or certain sea fish. How did they have the ability to change their colours to protect themselves from their predators? Did they suddenly decide to change their colours? It must have already been built into their DNA to have such an ability to achieve this. Where did that intelligence come from? Do other animals have the same ability built into their DNA, but are not aware of it?

At what point did the South American leaf insect suddenly decide, 'I know what! I'll make myself look like a leaf to disguise myself.'? Did it just happen overnight, or did it happen over millions of years? Or could it be that it was created like that? Evolutionists say that Evolution and Natural Selection have no thought, but just happen naturally. I just can't understand how the leaf insect decided it wanted to look like a leaf, or a stick insect to look like a stick. If it was not by Natural Selection, was it by chance or was it designed and created right from its conception into this wonderful world of ours?

There are so many other creatures that have incredible abilities and it's difficult to comprehend that they just happened by Natural Selection. Was all of the above just a chain of events that simply took its course, and we are just a part of this amazing process with no meaning, and one day everything will be destroyed and earth will just be a part of history? If that is the case, history itself would have no meaning, because history only exists in the minds of humans. So, is everything in this world meaningless?

If scientists have calculated it right that 99.9% of all life, including all creatures and plant life, has been destroyed by various historical catastrophes, then that logically means that our world as we know it today is now a new product of just 0.1% of everything past. So if you take the Evolution route, that means that nearly all living creatures and plant life must have evolved from the Jurassic period some 62 million years ago. Then, consider that there have been many other world ice ages since then. That means that Evolution and Natural Selection are absolutely incredible, as it takes millions of years for animals to evolve and mutate, 'if that mutation is at all possible', and that all life had to recreate itself with all new animals, insects, and plant life, in a limited space of time.

However, we could look at it in a different way. Perhaps 50% of the Jurassic life survived 'excluding the dinosaurs', and life evolved from there until the next catastrophe, then a percentage of that survived and so on and so on. That could be how it happened, rather than the 99.9% dying out all at the same time. But it is still a vast amount of evolving in a fairly short time in the world's history.

The last ice age was only 21 thousand years ago and lasted till about 11,000 years ago, during which time scientists believe that 75% of all large species of animals did not survive. And that could have been a lot more, including the woolly mammoths, the mastodons, sabre-toothed tigers and giant bears. If this is so, where did modern-day elephants and tigers come from? Where did the elephants evolve from? Are they a new creation? It's another one of life's mysteries.

So, did Evolution give us a completely new natural world? The above theory is quite plausible, but it does not explain the beginning of all life in any way. There must have been, in my opinion, some intelligence that kick-started it. And if so, did He just abandon us and let nature take its course?

Can you name just one species of animal that is in the process of evolving into a completely new species of animal, or any type of flower, fruit or plant? A rose is a rose. There is no animal that is a mixture of two animals, for instance, part sheep and part pig! Or any fruit, say part apple and part plum. One plant or animal might become extinct, but there is nothing in science that has ever proved to be part of one or part of another.

Continuing on from there, IF we are still in a process of evolving (and there is nothing to say we are still not evolving, as Evolution is a perpetual process), logic tells us that there must be a creature in this world of ours that is part one animal and part another. Every new creature that has been discovered has its own identity and has been given 'by us or God' a completely new name, for instance. If we found an animal that was part rabbit and part hare, we could call it a 'harebit', or a wasp that could make honey we could call a 'wasbee'. But as far as I know, there is no living creature that shares its name with another species. They are all unique, with their own DNA. Where is the mutation process? Geology has never unearthed anything to say otherwise.

The First Man!

How is it that if we go back three million years when chimpanzees roamed the earth, in a space of under a million years the first man appeared and came into existence? Did a primeval ape or chimpanzee first escape the ice ages and then mutate? Which animal did the chimpanzee mutate from then evolve into the first human, in the space of under a million years?

Looking for the missing link geographically and scientifically between chimpanzee and man, you have to consider what's known as the Last Common Ancestor (LCA).

There have been many evolutionary biologists, going back to Darwin and his colleague Thomas Huxley, who have been searching for the missing Evolution link between man and apes. But no matter how long they searched, and after investigating thousands of fossils, they have not found one fossil to link man to ape.

Perhaps they have not found a link because it does not exist. Even with the benefit of DNA, the evidence does not prove we are descended from apes. Maybe, if they continue to search for an answer within their evolutionary theories, and are trapped within those boundaries, they might be forced to think outside the box and even entertain the possibility that we were individually created and were not descendants and evolved from first living cells.

I will look at individual creation possibility in the paragraphs a few pages on, when we explore the Cambrian Explosion under 'Darwin's Dilemma'.

All living creatures have the ability to adapt into their surroundings, but there is no evidence that a creature could evolve into a new species. Only in mythology can you find the idea of part horse and part man, or a mermaid that is part human and part fish, or a creature part man and part lion. In the mythological figure of Pegasus, you can see only in man's mind that he would like to run as fast as a horse and fly like a bird. But that is all in his imagination.

Mutation. If you take a computer and put a random direction into its hard drive, would it not detect this as a virus and more than likely shut down? If you gave a box of cards numbering 1-100 to a toddler who has not yet learnt how to count, how likely would he or she be able to put them in exactly the right order? The first number would have to be one – a chance in a hundred – the next number being two. That is a chance of 100/100. Or if you had a padlock with a hundred combination numbers, how likely are you going to hit on the right combination? If it just had three combination numbers, you have a chance in one hundred to find it. What chance to win the lottery choosing just seven numbers?

This is the same for any living creature. If you put new information into its gene, unless this info is first of all compatible, but also beneficial for the animal to improve its life or change its function, then what good is it? Has the animal got the ability to throw out any negative information built into its gene, and just select, beneficial info? Natural Selection. If it selects just good information to enable it to change, why then does it never select bad information?

You could say if that did happen then we would have the strangest of creatures walking our planet. So we know that if mutation is real, it can only seek perfection. The ability to select, requires decision-making with at least basic intelligence, otherwise it must be completely random. If Natural Selection is random, it would take a vast period of time just to change some of the features of an animal, let alone change its DNA to become an entirely new species of animal. Mutation?

It is commonly believed that one day our world will be no more, whether by a natural catastrophe or by the other belief – by God coming again into our world and ending it all. My question is: Why has it not happened previously? OK, if 99% of all life is now extinct, it could so easily have been 100% on many occasions in our world's history. Did God come in another way, by sending Jesus just 2,000 years ago to warn us and to show us there is another world outside of this world of ours? Or is it going to end in a big bang, just like the Big Bang theory, i.e. Life started with the Big Bang, will it end the same way? That to me, is just meaningless. Maybe it is, as some people say, that man has got to have something to believe in, otherwise what is the point of life? If that is the case, so be it.

I prefer to carry on in my delusion, that is of course if it is a delusion. Or have I just stumbled on the truth which has been there all the time and we have not recognised it? Has it been staring us in the face all the time?

After weighing up all the possibilities and probabilities: Is God the biggest hangover? Or, is it Evolution with Natural Selection?

Our world is far more complex than just Evolution and Natural Selection. Our world was placed exactly the right distance from the sun, and tilted to exactly the right angle for life on earth to exist, with exactly the right atmosphere and eco-system. There are far too many co-incidences for us and this world just to have happened by pure chance, and the odds of it happening ever again are a billion, billion, billion-to-one. In fact, I would dare to say it is impossible. Even Professor Brian Cox stated in a 2019 TV documentary that if it did happen again, it would never be the same as our planet. He recognised that the only planet in our solar system that could sustain life is Titan, and that would be providing it took on vast changes to its eco-system.

Taking all of the above into consideration, no matter which way I turn I keep coming to the same conclusion: Our world must have some element of design in it! If this is true, who or what created it? I look at the beauty outside my bedroom window; was this created for you and me? And this leads me to the question: If it was created who made it? The three major religions in our world are followed by Muslims, Jews, and Christians, who all base their beliefs on the God of the Bible which states at the very beginning, that He made the world. What other explanation can there be?

Many of our best scientists, both alive and dead, have all come to the one conclusion that with all the theories – and many have an element of truth in them – this world could not just happen by chance, so it must have had a creator or a designer. Even Charles Darwin, in his latter life, had to concede that his life's work, although it had much truth in it, was based purely on theory. I also believe that many of our scientists of today, including Professor Brian Cox, will eventually come to the conclusion that our world is fearfully and wonderfully made, and there must be a designer!

Darwin's Dilemma

He was troubled over the absolute record in the historic geological appearance of a significant amount of entirely new animal creatures that suddenly appeared 'with no apparent ancestors' in the earth's sedimentary layers. This later became known as the Cambrian Explosion. Darwin sent a copy of his book, *Origin of the Species,* to one of the best known palaeontologists and scientists of that time – Swiss-born Louis Agassiz – hoping he could give an explanation to his dilemma. But regrettably, Agassiz concluded that the fossil record, particularly the record of the explosion of Cambrian animal life, posed an insuperable difficulty for Darwin's theory.

Darwin was hoping that one day science would give him an answer to his dilemma. However, science and geology have not to this day come up with an answer to this mystery. This event happened about 530 million years ago – approximately 300 million years before the Jurassic age.

Some of the best-known palaeontologists and biologists have investigated this sudden appearance of fossils, but none can provide an answer. The neo-Darwinian believers have also failed miserably to come up with a plausible theoretical answer, and many of their teachings on Evolution do not even mention the Cambrian Explosion.

With modern technology, this dilemma has made this old discovery even more complicated than ever before, and enhances the probability of Intelligent Design.

Well known philosopher Stephen C Meyer, in his book *Darwin's Doubt*, suggests this could be a divine intervention as one of the only plausible possibilities!

I realise the Cambrian Explosion causes an additional problem. If you believe completely and solely in Evolution, with no other belief for the reason for our existence, then the sudden and mysterious appearance of completely new life forms – without any previous ancestor – logically means there must have been some other external force or intelligence for their arrival on our planet. If you are a pure neo-evolutionist with no belief in intelligent design, then this could seriously challenge your atheistic beliefs.

With the latest technology, scientists have narrowed down the Cambrian Explosion to a period of 525 to 530 MYA, a period of 5 to 6 million years. This can be defined as just one minute in a 24-hour day, or 1/10 of one per cent of the earth's history, which is clearly illustrated in Stephen Meyers book, *Darwin's Doubt*. (p71-72).

My last thought on this chapter is that if scientists have quite conclusively proven that the first animals appeared on earth approximately 558 million years ago and the Cambrian Explosion happened just approximately 28 million years later, could they be one and the same event? And all animal life evolved from then? Or was it the first Garden of Eden event for animals, so to speak? Could there be other times, like when the first prehistoric man appeared on earth, about two-and-a-quarter million years ago? Evolution or Intelligent Design?

Chapter Three

Prehistoric Life

If we look at the word prehistoric, that suggests pre-history. OK, so when did history begin? Well, some of the first written records began when the early Jews wrote the first five books of the Bible, called the Pentateuch. There are many other early writings outside the Bible, from the early Egyptians and others, and all from the same part of the world from the same period of time. But they do not bring a constant history of a nation like the history of the Hebrews.

So, if we use the possibility that the time before the Bible had to be prehistoric, that does not include what is proven by geology and other scientific discoveries which could be defined as the history of the world. I know this will not be accepted by the Christian Creationist Church, but I will try to address this dilemma in a later chapter.

Let us first look at prehistoric animals. When we consider the recognised long, overall plan of things, the world began some 14 billion years ago after the so-called Big Bang theory. Scientists have calculated that animal life came into existence approximately 558 million years ago and a sudden influx of animals known as the Cambrian Explosion some 28 million years later. Then there was the introduction of dinosaurs, 200 to 230 MYA, and their existence lasted until 65 million years ago when they became extinct after the onslaught of an ice age. Or could it have been a meteorite colliding with our planet earth?

The first primitive man is thought to have come into existence about 2.4 million years ago, based on geographical findings. That is more than 60 million years after the Jurassic age, so if that is the case, how did prehistoric man come into existence? Did he evolve from apes? If so, where did the apes come from, and so on? Somehow both dinosaurs and prehistoric man came into existence and existed in different time spans; dinosaurs became extinct before man arrived on earth, as geologists haven't found any evidence of dinosaur bones in the same deposit as human bones. We are 'post prehistoric man', so from now on I will call us 'modern man'! Are we descendants of prehistoric man, just 6,000 years ago? It makes one wonder if the creator of the world was not experimenting with types of animals and men in anticipation of a planned new life on earth? That is just a thought!

Geologists have calculated that the last ice age started a mere 21,000 years ago and ended 11,500 years ago. Was this over the whole earth, or was it just in the western hemisphere? Could that have killed off all life? It makes me wonder if that left the world void of all life, and then this new civilised life as we know it today – being modern man, not a life just living on instinct – was recreated some 6,000 years ago with Adam. Whatever happened 6,000 years ago, it was something very special.

Why is modern man so different to animals in so many different ways? If we look at chimpanzees, the world view is that they are our nearest relatives, which is quite understandable and plausible. But the discovery of DNA shows that our DNA is different from theirs. And though there may be many similarities, it is not possible for them to evolve from monkey to man, because DNA cannot mutate in such a short space of time. If it was at all possible, it would take millions of years: a cat is a cat, and a dog is a dog. Therefore, if we take this analogy, man is a separate being.

There is also a logical reason for prehistoric man and us. First, geologists say the last ice age was some 21,000 years ago, lasted about 10,000 years, and probably wiped out most, if not all, living land creatures, certainly in Europe and a large part of Asia and Africa. There is also some evidence it happened in the Americas, as worms became extinct, though we cannot be sure of this, and they are vital in the irrigation of crops.

It is believed that Homo sapiens like modern man started their existence about 200-230 MYA, and throughout that time they lived primitive lives and never developed like modern man, who within the space of 6000 years have become so developed that they have out surpassed anything before. If we have established that man is a different creature to any other, it begs the question: what had changed with this new man who, after all, still had more or less the same size brain? In less than 2,000 years he went from living in caves to building palaces and pyramids, to living in houses. If you take the analogy of an inch for every thousand years, 6,000 would just be six inches, but the first prehistoric man lived over 180 feet ago.

Anthropologists believe they have discovered at least five different species of man going back through time, each with their own distinctive DNA. Modern man (Homo sapiens), with their DNA, being one of them. They have a percentage of the other species' DNA similar to ours, which throws up a very interesting concept. If they did not evolve from any previous animal or man, and they are a different species from chimpanzees, then could the creator have created them separately with the ability to interbreed? Thus, if He did this, then it's possible that He did this at different times. Which also brings up the questions of extinction by ice ages. Was all mankind destroyed at any one given time, or did different pockets of man survive?

When we look at the Bible's account of history and the flood with Noah about 4,500 years ago, did that engulf the whole world? Or was it isolated to the Middle East, Europe, and possibly Asia, as there are legends of a great flood around the same time in China? Taking the above into consideration, it is possible that different species of man have been created independently at different times and did not just evolve from one to another. But did they have the ability to interbreed?

The creator, through Adam, could have made a new subspecies of man. I've tried to incorporate most of the possibilities and probabilities, and I'm sure as time goes on there will be more apotheosis and theories thrown into the melting pot. I have not attempted to delve into the different species of man, like the Neanderthals, Denisovans, Archaic Africans, or others, as we are uncovering more and more about the history of man's evolution.

In the next chapters I will give reasons why the new man, through Adam, was completely different from anything before.

It's interesting to note that most geological finds go back to the stone age, and seem to go back no further than 6,000 years. These include Stonehenge 5,000 years ago, Maiden Castle up to about 6,000 years, Skara Brea and the Ness of Brodgar in the Orkneys 5,200 years ago. It is as if everything seemed to kick start from 4,000BC, as if there was a clear gap of no or little human activity between 6,000 years ago till 21,000 years ago, with very few exceptions, certainly in Europe and Asia.

It's strange, but around 2,200BC and 2,500BC the Orkney civilisations suddenly came to an end. But why? Could it have been a huge flood? This was at the same time as the biblical flood of Noah, at 2,356BC.

A resume of the last three chapters. And prologue to the next part of this book

First, I would like to tell you a story, and it goes like this.

A letter was sent out to the world's top five physicists, the top five scientists and the top five inventors. They were chosen as the best of best in all the world in their subject. And they were invited to a creative seminar of the utmost importance! The name of the campus was called the UNIVERSITY OF LIFE, and the letter was signed by the Head Principal. They all accepted the invitation, keen to meet like-minded people, and hoping to learn more.

When they arrived, they were treated very well with the best of everything, and after a while, were led into a conference room where the principal told them of their challenge. He told them: 'I want you to create a single blade of grass. With the ability to reproduce.'

Their immediate response was one of insult. 'You brought us from all over the world just to make a simple blade of grass!' But they were more upbeat when they told the prize would be a million pounds, given to each of them to give to any charity or good cause of their choice... but only if the task was completed.

Well, they were enjoying one another's company so much that they agreed. They were then told the university would assist them by having certain aids to help them get started. They would have a laboratory which would give them the *perfect climate* for the grass to grow and germinate; it would have as much *water* as was needed; and *soil* perfect for growing. In addition, any other scientific equipment ever invented would be provided to assist them, including computers. Then they could call on any other person either still alive or someone from the past, like Louis Pasteur, Alexander Fleming, Archimedes, or whoever, to help. And last of all, they would have as much time as they wanted to complete the task.

They were all very keen to get started, but soon began to realise it was not as easy as they had thought. 'Ah! We need a plan. How did the world begin? Nearly all of us have written books or made documentaries on how we thought/believed it all started. Right! When our planet earth was just in the *right position in the Universe* to give us the *perfect ecosystem* we required, that was after the Big Bang. There was *water*, and earthquakes happened, then there were *nutrients and oxygen joined together to create the first cell*. How? Never mind, let's skip that for now; let's look at the first cell that is said started in the sea, then it grew a head and then a tail. Well, we don't know why, but do we know how it really happened. Then, from there it grew into all living creatures as it evolved in different ways and adapted itself to suit its need to survive!

'Hey, one minute, weren't we meant to make a blade of grass? Well, perhaps from that original pea soup, fungus grew and that evolved into grass. But where does *DNA* come into all of this? Well, we know all living things have *DNA*, including a blade of grass, but how do we create a *DNA*? We just do not know.

'Let's go back to the principal and ask for more help and more time to invent a way of creating a *DNA*, and fit it into the structure of a blade of grass. Ah, but that time is not now, and it will be some time in the future.'

They put their findings to the principal, but he said, 'You have failed. With all your books, theories, and inventions, you have failed to find the meaning of life. Many of you still believe it all started by a chance happening and a big bang.' Then he lifted a simple blade of grass up in his hand and stared at it intensely and said, 'This simple blade of grass has more intelligence written in its DNA than all the intelligence of your whole group put together, past, present, and future.'

Then he ushered them forward and put a book in front of them, which they immediately thought was to show them the secrets of life. On the front of the book was written in gold letters, THE BOOK OF LIFE. Out of curiosity, they opened it, and on the first page was written: These are those who passed their Degree with Honours.

As they turned the pages, they saw many famous names – some they knew. They were surprised to see Charles Darwin's name, and they asked the principal, 'Why?'

He told them, 'Charles was sorry he ever wrote the Origins of the Species, as it was based only on theory, not on fact, and many people believed it to be true and it became their religion, so to speak. Towards the end of his life on earth, he realised his devoted wife, who was a Christian, was right all along, and that this world of ours was far too complicated to happen by pure chance.'

The next name after Charles Darwin was a little boy named Tommy, who died aged ten, and a little girl named Jane, aged eight. They asked the principal, 'Why are these in the book? What do they know about life at such a young age?'

The principal said, 'They loved the creation all around them, and thanked God every time they went to bed. They died trying to save their little baby brother from a fire in their house.'

They looked further into the book, and there were the names of an elderly Jewish couple. They asked, 'Why?'

The principal replied, 'This couple always throughout their life kept faith in God, and even when they were executed in the gas chambers by the Nazis in World War Two, they never stopped loving God and believing in Him to save them.'

They then asked, 'If He – that is God – is real, why does he not show us who He is and how He created our world?'

The principal told them, 'He showed this on all of His creation, then He sent His son Jesus to tell you and show you who He was. He even told you hundreds of years earlier what to look out for in Jesus' coming, in the prophecies in the Old Testament Bible, but you did not listen.'

The group of intellectuals was left confounded.

In the above story, the words I've written in italic are to highlight some of the major elements needed to create and sustain life.

In this story, I'm not saying that Charles Darwin became a Christian on his death bed, in line with a certain myth; he was neither an outright atheist or a born-again believer, but I would love to be able to talk to him if and when we get to heaven! Trusting we will both be there.

Beginning of the World Theory

The World was created, or somehow came into existence: Time! Space! Matter! What happened? The Big Bang theory? There are so many elements that when considered make it too incredible to think that it was created into a perfect situation for this to have just happened by chance! Or did it? Then, when the time was right, the eco-system had established itself, and the stars were in the right position, so to speak, life sprang into action. Was it from a pea soup, or was it created? Were creatures created or did they slowly evolve from a tiny cell or cells? Where did they come from, with a vast amount of DNA built into them? Why was there a sudden explosion of creatures 530 million years ago, with no ancestors? The Cambrian Explosion. Was the Charles Darwin theory credible? Then we moved into the Jurassic period, 230 MYA, and they lived for over 150 million years up to 65 million years ago.

Going on from there were several ice ages, but animal life continued to thrive; and evolve? Till about two-and-a-quarter million years ago, when the first recognised man appeared. Then more modern prehistoric man, Homo sapiens, came into existence and lived between 200,000-25,000 years ago. Geology is a bit sketchy on these dates, and it is believed they were a sub-species of the Neanderthals , though there is believed to have been an ice age between 70 and 90,000 years ago, and man survived. The next ice age was 21,000 years ago and lasted 10,000 years, which takes us up to 9,000BC; was this a global ice age? Did some pockets of people and animals survive somewhere on our planet? We know the woolly mammoth and the sabre tooth tiger became extinct around 10,000 years ago, and the elephant and modern-day tigers appeared on earth sometime after that.

However, I believe there is enough evidence to say modern man started with Adam, in 4,000BC. This modern man, in a small space of under 2,000 years, had gone from living in caves for hundreds of thousands of years to living in palaces, and had an ability to communicate in a civilised language. This sudden transition began with Adam; God chose to create him in His own image. It is said in the Bible that God breathed His Spirit into him. There is something almost magical in what happened when God created this new kind of man in Adam; that is, if you can accept the possibility that there is a creator God. You may think: *No, Adam and Eve were just a fairy tale made up out of someone's imagination, and the Hebrew nation's history is built on a pack of lies.* Well, in the next chapters, I hope the evidence supporting the Adam and Eve story demands a verdict that it is beyond reasonable doubt, and the first accounts of man are in fact actually a true account of the beginning of modern man.

So far, I have explored the start of life on earth and how the history of the world came into existence, right up to 6,000 years ago, and continuing into the beginning of a completely new era that is, if true, so incredible that it will completely change our views on who we are, and give a fresh understanding and meaning to what life is all about!

Fiction or non-fiction?

If you are of the mind that the world started with the Big Bang, and life is based on Natural Selection and there is no designer or creator, then this book would end here. As there is nothing to say if and when this world of ours would end, the Bible is the main chronological record of the history of modern man. You may believe the Bible is mainly "make believe", but I hope to show that a large part of this book – even if you don't believe in its authenticity –is correct and the best historical record of the last 6,000 years that we have. I will also look at the evidence that backs the Bible's credibility, and show it is not just a book of stories!

We will explore why the man in Adam and Eve were so different to any living human before. If you are still of the opinion that the world started by a primeval soup or something like that and we just evolved from there, then what I say in the next chapters may have little or no meaning. But if you believe in the possibility that there is a design to our world, then it naturally follows there must be a designer and creator, so let this be the challenge: To find more about Him! This can be discovered partly within ourselves, because I believe we have many of His characteristics and spiritual elements written in our DNA, and in many ways we have been made in His image. Examine the Bible to see what it says about Him.

I've written what I believe is an ongoing pattern that not only conforms to science and geology but also to the Bible. I'm asking: Is this Creator God one of the same – the God of all creation and the God of the Hebrews? And I've written how they are intricately connected, and that the Bible is a correct historical book that can be relied on, and not fantasy.

There are many reasons why this book has credibility. If we started a new era in man 6,000 years ago, then why – if you do the calculation and take the genealogy of the Hebrews as recorded in the Bible – does it go back almost exactly 6,000 years? This makes it either fabrication, a load of coincidences, or a correct record of our historical past. Another oddity is that we have always recorded time by the seasons of the year and by the hours of the day. But the earliest records of the seven days of the week started in the Bible!

From now on, this book will bring a third element to our existence – that being our spirit; the others being mind and body.

Throughout my years of being a Christian, I've experienced many events that cannot be explained logically or scientifically. Some people may say they are all in the mind, others might say there is another reason or even magic, whatever that might mean. Others say they are just coincidences, but my relationship with God is real and so it is with billions of other people in every part of the world. It is where our spirit is in close communication with the Spirit of God, and I hope if you have not already experienced this relationship that this book will guide your mind to seek what billions of others have discovered, in nearly every part of the world.

I'm sure if you choose this new understanding to life, you will never be the same. It has the ability to change your outlook on life in a multitude of ways for the better, and so much more! It will also show you the reason for your existence and that there is hope in a life after death! By releasing you from any doubt or fear, it's like finding a lost treasure map and finding the treasure at the end of the rainbow or the missing part in our DNA. I do hope that no matter who you are and whatever you believe, you will continue to read the rest of this book, even if you come to a totally different idea of what we are and where we came from.

If you have come to the conclusion that there is design to our world and that it has been created, then by whom or what, if it is not the God of the Jews, the Muslims, and the Christians? What other choices do we have?

    You may believe there is another explanation, that there is a creator of this world but not the God of the Jews, Muslims, and Christians. If so, I would love to hear from you.

Chapter Four

The Birth of Human Civilisation and the Uniqueness of Modern Man in the Hebrews

Before we can begin, I would like to clarify who modern man is and what he is not. First, he is not the same as prehistoric man I previously put forward the reasoning that 6,000 years ago, this new man was so different, so much more developed intellectually, that he out-surpassed anything previously. It begs the question: What had changed with this new man, when he still had more or less the same size of brain? Yet in less than 2,000 years men were building palaces and pyramids and living in houses. What changed in modern man? It is believed that prehistoric man not only lived mainly in caves – hence cavemen – but they lived like every other animal or creature and were driven by the survival of the fittest, without the sense of reason or knowledge of good and evil. This new understanding only began with the start of modern man. And as far as we know, the Bible is our first record of man having an audible language, with the ability to reason.

The first concept of history is in the Bible. That is, if in fact, the Bible is our earliest reference book. It dates the beginning of modern history as starting 6,000 years ago. The genealogy of the Jewish nation is calculated to prove this timescale in an illustration later in this book. It also presumes that the Bible is not just a book of stories made up by the Jewish nation some three-and-a-half thousand years ago.

When did History and Time begin? It only began when we had the capacity to comprehend them, so where did this ability to comprehend come from? Thin air? The first record of time is recorded in the Bible as being seven days. Genesis, Chapter 2, verses 1-3: 'So the heavens and the earth and everything in them was created. On the seventh day God had finished his work, of creation, so He rested from his work. And God blessed the seventh day and declared it holy, because it was the day when He rested from all his work of creation.'

I will tackle the thought that God rested on the seventh day later.

Words! First, we must have a language before we can record events and create pictures or signs for the meaning of each word.

History is only history if we can remember it and ultimately write it down to preserve the thought, be it a genuine event that can be proven or a chance happening.

I will look at what history is in the last chapter in the book (Chapter 12).

Prehistoric – that is pre-6,000 years ago – was completely reliant on what each individual could remember, then have the ability to record such events particularly without the written word. It would be very limiting on what they could remember, for instance, how to get from A to B without a map. Also, without an audible language, prehistoric man's capacity must have been very limiting; deaf and blind people today can understand such difficulties.

It was not until we could communicate with languages and the written word came part of everyday life with the ability to record events that we could create civilised communities. Civilisations start when culture becomes so advanced that, through education, it creates a stage of social and cultural growth to develop an organised community, with a set of rules and laws and order for the benefit of all.

We needed these building blocks to create civilisations that had the ability to communicate and commune with one another in harmony. They also found that by living in communities they could not only protect one another, but they could find better ways of producing food, finding water, and increasing their population in a secure and safe environment.

The earliest known communities were thought to be in Mesopotamia, about 3000BC. The same era and area that the first recognised written languages were recorded.

If the Bible is our first record of living thought… It was written in the first spoken audible language, Hebrew, and the first and only book that records the complete history of a nation from the beginning of their family. Adam and Eve, at the beginning of their conception, were just two people that gave birth to the whole Hebrew Nation, and I dare say that probably includes you and me.

If it is the first history book that modern civilisation based its culture and laws on… Its history shows it came from the oldest civilisation going back 6,000 years up to 2,000 years ago. Its translations have been word perfect in all the known found ancient manuscripts, and that includes the Dead Sea Scrolls.

If the Bible was just a made-up selection of myths and stories, then the whole Jewish nation are living in a pretence, and their credence and genealogy would be a fabrication. That would also bring into doubt the existence of Jesus, the validity of all Christianity, and the foundation of what Christians believe – and also what Muslims believe in, as their foundations go back to Abraham and the first chapters of the Bible.

So, there is no other book to compare with it; the Bible is unique in every way. In addition to all this, it shows that we are not just part of an evolutionary sequence but a new creation in modern man, born from a real and living Creator God, starting with Adam!

Why was Adam so different from any other creature or prehistoric man? Adam was born in the image of God, reflecting His nature; man had God's Spirit blowing through his whole being. If this is true, what does it mean? Was Adam God's perfect creation? Far superior to any other creature that had ever been before, he not only had God's nature (Genesis, Chapter 1, verse 26), he also had His creative ability embedded and carved in his body and soul. If God made Adam, He would have known him so well and been fully aware of all his needs, Adam being part God yet still human. So He gave Adam a partner – a woman to care for him, and they would love one another as God loved them. This love for them was far deeper, stronger, and wider than any other creature that existed before.

What is love? Did love exist in prehistoric men? God's love for Adam and Eve was far more intimate than anyone could imagine, and because He loved them so much He gave them free will, with the option to do good and please Him, or wrong! Perfect love is to completely trust.

There's a fairy tale, The Prince and the Servant Girl: Once upon a time, there was a handsome prince, and in his palace there was a pretty servant girl. But the prince took no notice of her, so one day she met an old lady who knew her plight, in that she loved the prince, so she gave her a magic potion. All the girl had to do was to put a single drop in his drink every day and the prince would fall in love with her. So that is what happened. The prince fell in love with her and they got married, but then she thought, *I do not want my husband to love me because of some magic potion. I want him to love me because of me.* So one day she took a chance and stopped giving him the potion, and found out that he loved her even more.

God wants us to love Him freely. He did not make us like robots; He gave us freedom of choice.

God knew what He was doing by giving Adam and Eve this freedom. He knew they would be tempted to do wrong, and He knew this would be a curse on all mankind for time to come. But ultimately, He also knew that through much suffering, man would turn in his need and come back to their creator, the one who loved them through good and bad, right from the beginning. But why would the creator of our world ever embark on creating a being that would be able to possess such evil that was completely opposite to his own nature? One of the answers could be that He wanted to increase his family! Or hunger for an increased love for Him and all he had created. Or you could believe it was to populate his heavenly kingdom. Another reason could be that He wanted to prove good would always triumph over evil. Would it have been better for God to do nothing? Well, if He had done nothing, then you could certainly say there would be no evil in this world of ours, as you need humans through Adam for evil to exist, as pre-Adam was back to survival of the fittest with no moral thought.

Allowing evil to exist allowed good, love, to exist even more! If there is no evil, good would have no meaning. I will talk about evil, what it is, and the Devil later.

Many of the Jews in the Old Testament Bible turned against God. Even though God cared for them, often in miraculous ways, they turned their backs to Him. God's saints and prophets turned in their despair back to Him, and He also made a perfect way back to Himself by allowing His only son Jesus to suffer, die, and make a perfect way to rid us of all wrongs and come back to Him. God loved us as much. If this was not so, He would not have allowed His only son Jesus to die in our place. How wonderful is that!

As long as we turn and accept Jesus as the son of God, that gives us a free ticket to heaven, because Jesus paid the price. Love will always defeat evil. We also have the promise of receiving His Spirit, the Holy Spirit, in our lives, the seal that is eternal, that guarantee of a place in heaven after our physical death. If God created us, He wants us to be a part of and join Him in His eternal home.

History starts from Genesis; that is what we use as the benchmark of our History, as there is nothing recorded before Adam! If this is the totality of our history, there can be nothing that is prehistoric, because everything before then is known and prehistoric, and would not exist. But we know that this not true. We have discovered through science, physics, and biology, that many things existed millions of years before Adam and Eve. Or did they? Logic tells us there must be something that brought our world as we know it into being – whether it was as it was described in Genesis, or whether we have prehistoric?

I believe God chose to create in us-, a wonderful being, far better, far more beautiful, and with so much ability to do things that nothing before could ever do. It is as if He had decided to paint the most beautiful painting ever, with everything in creation incorporated in it, but it was not complete until He signed that painting with His signature, putting His seal of approval, and breathing His Spirit within man. I call this the missing link between God and man. And this is the most, most amazing event in all creation. We are the Creator's pride and joy. Without His Spirit living inside us, we would only have the same existence as prehistoric man or all other living creatures. If you can believe it, does that not excite and amaze you? That He loves you so much He sent His only son Jesus into this world to die, 'so that He could restore that same Spirit in you, back to Himself'. The missing link! Jesus was that seal; His signature on your life. I trust you can bear with me and I will later explain what part Jesus plays in the overall plan. Was this His plan before the beginning of time? Whenever time began?

Another aspect of the uniqueness of man can be seen in God's love for the Jewish race – God's chosen people. Antisemitism is born out of a disbelief in God,

The Jews even to this day are an amazing race of people. They can boast some of the greatest artists, actors, film producers, businesspeople, bankers, and jewellers, and have more millionaires per population than any other race. That was one reason why Hitler wanted to eradicate them, as he knew deep down that the Jews were a far more superior race to his.

They are also the only known race of people that have established themselves back into their own country after thousands of years of roaming the earth. And no matter how hard other countries try, they cannot move them from Israel, or destroy them. They are also the only race of people that has been dispersed throughout the world but still retained its identity. If you look at any other race that went outside the bloodline of Abraham, to see what they have achieved, it is nothing compared to the Jewish race.

Somehow, one could say, they have almost miraculously survived hate, persecution, genocide, eviction from their homeland by the Romans, and scattered to the four corners of the globe, but still retained their identity. One of the main reasons for this is that they have stayed true to their faith and the laws given to them by God through Moses in the Bible, by not marrying outside their faith. If we look deeper into this, and if we accept the possibility there is some evidence in the Bible, it will all make sense.

First, if the initial modern man made was Adam, and his creator God made him special and different to anything else before, God put part of Himself in Adam. It was written in Adam's DNA and carried on from there, through to Noah, still keeping the lifeline pure, and after the flood again keeping a strong bloodline through to Moses, when the law was made to not marry outside the race. There was a strong reason for this, in that their DNA from Adam should remain pure and not be contaminated from outsiders. This is why the genealogy is so important to them. If there was anything in the theory of other races existing pre-Adam, the bloodline could be affected.

I don't know of any other race, country, or clan of people that has kept its genealogy back to the beginning of conception; it is only in recent years that people have tried to trace their family tree. With the Jews, it was imperative they kept their DNA pure, so when it came to Jesus being born, Mary's unbroken bloodline was recorded directly back through the generation to David, Moses, Abraham, Noah, and ultimately Adam.

If you take a bottle of water and add a red dye or any other colour, then it is diluted with more water, eventually the dye will fade away. This is why it was imperative they kept the bloodline pure, so that the uniqueness of Adam's DNA would remain. It would be good if a study could be made with the Jewish race of their DNA, particularly with those who have kept as far as could be established a close bloodline going back through the centuries to Adam. No doubt there would have been some infiltration into their DNA.

There is much more one could say about the uniqueness of man, but hopefully this book will be a testament to how amazing we are and how we are fearfully and wonderfully made.

Apes can ape man, but they have not got the ability to create things like paintings and music or invent cars, aeroplanes, computers, and so on. Unlike modern man, they still live by the basic instinct to survive – much the same as prehistoric man.

How unique is modern man!

Chapter Five

Who were Adam and Eve?

It is commonly said that Adam was the first man and Eve the first woman. In some ways that is right, but in other ways it is not. In the previous chapters, I believe I have shown enough evidence to prove there were 'pre-Adam' men and woman. If so, why are they so different?

At their creation, Adam and Eve did not have any knowledge about anything apart from that they were created from the father God, and only knew what their creator father told them. With no earthly mother or father, that must have been so strange not being born. I cannot even imagine how that made them feel. Did they understand they were the first born? Or made people, and to know God as their father?

Were they created as babies, children, or young persons? Adam and Eve could not have been influenced by anything outside, but only by what God told them. How would you tell your child about prehistoric life at a young age? With no mums and dads, aunts or uncles, can you consider putting yourself in that situation, and just relying on what you were told by your father – sorry, heavenly father – with absolutely no knowledge and influence from anything else?

It must have been so strange for Eve, having no mother to talk to and share her feelings. But if everything was rosy in the garden, so to speak, she would have had no worries anyway – until they were thrown out of the garden, of course. But how would she have felt giving birth to her first baby? And experiencing pain – would you question things or just accept that as fact? So, Adam would have no reason to question what his father was telling him, particularly as there was no-one to tell him anything different.

The timescale from when Adam was born 'made' to the time of their first born was 130 years. There are no records of what happened, of how long Adam walked in the garden with God before God made Eve, or how long it was before they were thrown out of the garden. And Eve, how did she feel having no earthly mother or father, no childhood, instantly being a woman? Again, their children, Cain and Abel, had no aunts or uncles to tell them anything else than what their mum and dad told them. I can't imagine just how that felt.

If you also consider that God spoke in words to Adam, that would be the creation of the first language, one which Adam would have used to teach his children, and also passed on as the first complete language, using more than one syllable. Prehistoric man, as far as we know, would not have possessed a sophisticated language as Adam's family did. Why did this happen in such a short space of time? This just adds to the story of Adam being true.

Then man thought he could be clever and build a tall tower to reach God (Genesis 11, v1-9). This displeased God, so He dispersed the people throughout the then known world and gave them different languages. It is quite a phenomenon that different languages separate most countries even to this day. The Hebrew language is believed to have been in used before 1200BC, and is the same or similar language that Moses used to write the Pentateuch – the first five books of the Bible.

I can't imagine God sitting down with Adam or walking with him in the garden and saying, 'This is only conjecture. Let me tell you how I created the Universe' in basic details, or in any scientific way. I can imagine Him only telling Adam, 'as a father tells his son', the basics, nothing about the world's past or what He had created before. If He had told Adam in any more complex details, it would have been too much information! And of no use.

The early Jews had a much simpler life; they didn't have the scientific knowledge that we possess which allows us to understand prehistoric.

Adam passed down through the line of his family the beginning of what he believed was the start of life on earth. If these were not the facts as Adam knew them, then this would be probably considered the greatest hoax ever! There was no reason for Adam to lie or say anything else, because his children would have been part of the hoax, as there was no-one else alive with them at that time to tell them anything different. If you follow on from there, that would also mean the whole of the Bible, the history of the Hebrews, the Muslims, the stories of Jesus and so much more, would have been based on a lie, just a fabrication! And life as we know it would, in many ways, be meaningless. The stories given to us by Adam, of the first history of modern man, are so vital in our beliefs about who God is! We would not even have seven days a week!

When God told Adam that the world was made in six days, and on the seventh day He rested, this was to give Adam a model for how he should live his life. This established the first timescale for life, i.e. why a week was made up of seven days. Why not six days, or even eight, nine or ten days? It's strange that has never been changed throughout time!! And the whole world still uses it.

A day was only a period of time, and if God created everything, time was just part of His creation. A day in God's time could be anything; it could be a day as we know it, or a thousand years, or much more. And is there something else outside time as we know it? God being God could create anything or everything in one earthly day. God being the same yesterday, today, and always, He does not need to rest.

But us humans, He knew we needed to rest, so that is why He also created night, so we could sleep and rest our bodies, and the same on at least one day in every seven.

God created time! He does not need to rest at any time, let alone at night. Anyway, night is at a different time, depending on what part of the hemisphere we live in. It is always night somewhere! Does that mean He is asleep in Australia but awake in England at the same time? No, He is omnipotent. He is in all places at the same time. He is not limited to time and space.

God knew what was best for man, and wanted him to rest after six days' labour. He also wanted to give him time to contemplate the beauty around him and devote time to give Him thanks for all that we have been given. He also created day and night, knowing man needed to sleep.

We read in the Bible that when God made the world, He created man in his own image! And He breathed His nature and His spirit into him. The Bible also speaks of how He created the rest of the world in five days, but I will try to answer that question later. What did God mean when He created man in his own image? I don't believe He meant a physical image, although man has often painted God with long, shiny, white, flowing beard, full of wisdom, sitting on a throne. No, His image is a spiritual one... whatever that means,

So, why was the man in Adam so different to anything before? This Adam walked upright, and the spirit God breathed in him was part of Himself, with a whole range of creative qualities that were not given to any previous man or creature. You may ask, what are these? Well, all previous animals and men existed on the survival of the fittest basis; men lived in caves and would fight for food to survive. Whereas this new man had a built-in reason of right and wrong. He also has a creative force, spirit in him, and is able to create beautiful sculptures, paintings, music, and so much more. We know prehistoric man did paint cave drawings, but they were copying the world around them, not being inventive and their efforts not based on reason. Apes will ape things, but this is not being creative; their DNA is different to man. This new man is so much more.

Why is the time 6,000 years ago so vital to a new start of life on earth?

It is not just vital to this book. It is vital to a new start of life on earth. You could say it is the new beginning and the end chapter in the earth's history.

The first chapter, or part of this book, was from the creation of the Universe including Time, Space, and Matter, up to 6,000 years ago, and our new man goes back from this day in time to 4,000BC. Let me explain! When we talk about a man or woman, we say they have a soul as part of their person. What is this thing we call soul? In definition, everything that lives has a soul. It defines who we are, whether it is a worm in the ground, a wild cat, or in fact any living creature. Some might say the Soul only exists in humans, not in animals. Could they be right? I cannot prove it one way or another.

What is the Soul? First, it is that element in us that science cannot explain, it is the same as the Spirit that lives in us; you cannot touch them and you cannot see them, but we know they live in us. Scientists cannot define them, as they are that God element in our lives that is incomprehensible. It is the same way you cannot see God or touch Him. If you are a pure evolutionist, can your soul, your spirit, and God be real? Do they really exist as parts of a human being? Or have we just invented them to describe something we cannot see? Science, physics, and biology cannot explain the soul – it is outside human understanding.

I believe we have two parts to our soul. The first part is the same as any other living reproductive creature that could include the tiniest cell; it is driven by a quest for self-preservation, and that includes prehistoric man. However, modern man has a second part to his soul, which was imparted to him 6,000 years ago when God entrusted us with part of Himself. You could say that part of God's creative soul is living in us. How can I be so sure that this is an actual fact? It is the whole basis on which this book has been written. Before 6,000 years ago, man did not have the ability to write, nor as far as we know had an audible language. As I have spoken about earlier in this chapter, he did not have any comprehension of right and wrong, science, geology, physics, or history. And I will stress, the Bible being our oldest history book just by chance started in 4,000BC. These all give credence and stand as evidence of a new era, a new beginning,

We have been given a new ability to create amazing things, but I am sad to say that also includes the ability to destroy our world. That ability is written in the second part of our soul. Another part of this new dimension in our soul is the ability and quest to comprehend what is its source and a desire to reconnect to it. When we pray, who are we praying to? We can't see God and we cannot touch or audibly hear Him, but there is something in our soul that tells us He is listening, He is real and cares for us. Are we being stupid in believing there is someone or something we cannot touch or see? Or, is it that hidden part in us reaching out to the source of our creation to the creator? The Bible tells us about how God made us in His image! What changed 6,000 years ago?

God breathed His Spirit into man's soul? Was this the event that started the creation of the first known civilisations? There were hundreds of thousands of years with prehistoric man, and they all lived by instinct and survival of the fittest, and lived mainly in caves. Then 6000 years ago, something amazing happened. Within the space of just over 1,000 years, humans went from living in caves to living in houses and building palaces; not only that, but they had languages to communicate and man had also developed writing skills. Why did this happen just then? This was the very time that the Bible said the beginning of the then known world started with Adam and Eve! So, did the Jewish nation have it right all along, with this very simplistic concept?

Whether you're an atheist, agnostic, or a God-believing person, there is so much evidence that demands a verdict!

What is the difference between our soul and spirit? If we look at a car, the make and model describes what it is. So, that is who you are: you are a human being, and what your name is identifies you as being different from anybody or anything else, like your fingerprint or DNA. But it is the fuel we put into the car (your spirit) that determines where and how far it can go, as it is with man to have the energy and ability to go and do amazing things. Whether it is your first step as a baby, or a desire to climb Mount Everest, that is your spirit, that is what drives you on.

What is the Holy Spirit? I will talk about that later in the book.

I would ask some of the great scientists and biologists of our day, like Sir David Attenborough, Professor Brian Cox, Richard Dawkins, etc, if they have a soul. And if so, I'd challenge them to describe it to me in scientific terms.

Chapter Six

Is the Bible credible?

How do you write about the most incredible book ever written? It stands alone. It is not only our greatest history book ever to be written, but it is the first and best book to show who the creator of our world really is. There is no other; it is unique in so many different ways.

There have been a multitude of books written on the Bible, so I just want to concentrate on the evidence for some of its authenticities.

IN THE BEGINNING. Is there a beginning?

The Bible is the beginning of our history as known, told by Adam. That excludes prehistoric creation, as Adam would have had no knowledge of anything before.

The Bible gives the date 3,950BC as the starting date for modern history, from the creation of Adam and Eve. Others say 4,002BC. For my calculations, it does not matter which one is correct, as the difference between events remains the same. That being 52 years, it could be that God created Adam as a baby. Or was he made a grown-up man? However, the Bible does give us the age of how old he was when he died – at the ripe old age of 930 years.

The first thoughts ever to be recorded are in the Bible from Adam, apart from cavemen drawings and other areas of thought depicted in the tools the cavemen made, this being prior to Adam. But it was not until 6,000 years ago that the first recorded thoughts were established, as recorded in the Bible. It is quite amazing to think that as far as we know there is no evidence of creative thought recorded before Adam!

Was the Bible the first recorded history book? I think so. There are many historians who would say differently, but I would like to see their evidence.

What I'm trying to establish logically is from Adam to Moses, the probabilities that the facts about the story of creation written through Adam in the Bible, are true. In other words, is there a direct story line from Adam to Moses? Was it was passed on mainly by word of mouth? And is it credible?

The first five chapters of the Bible, written by Moses, contain immense detail recording the first 2,500 years history of the Jews. These books are some of the earliest recorded writings, written by his scribes, approximately 1,500 years BC. That leaves us with the question: How did they remember and recall all the previous 2,500 years of history? Well, we can only surmise that they relied mainly on their memories and passed them down from father to son, until they invented the written word. Again, if you think that for the first 2,000 years their history had to be passed down from Adam to his grandchildren and onto Noah, who was born 126 years after Adam died, at that time there were still five generation of Adam's children still alive. So it is highly probable that Noah received the story of creation almost first-hand. If we carry on from there, Noah was still alive when Abraham was born and was a young man. Noah died when Abraham was about 46 years old, so I can imagine Noah talking to Abraham and telling him all about the flood, and also going back to his forefathers and to Adam.

It is probable that early biblical men had other written recorded events prior to Moses. Archaeologists have found countless fragments of writings dating back from about

2,500BC. So, in a space of only three to four generations of people and over 2,100 years, Adam passed their history on to his children, who passed it to Noah, and Noah to Abraham, then Abraham to Moses.

The invention of writing had to be established and available at the time of Moses, otherwise he would not have been able to use it to write the history of the Jews. Moses had scribes to write these first accounts. It makes me wonder why the Jewish nation were so emphatic in recording the exact records of their genealogy?

The genealogy of Hebrews (Jews)

Event	Date BC from Creation	Age of first son	Age of Adam to Event	Age when died	Year BC when Died	Bible ref Chap & verse
Creation Adam	4002	130		930	3072	Genesis 5:3
Birth of Seth	3872	105	130	912	2960	Genesis 5:6
Birth of Enosh	3767	90	235	905	2862	Genesis 5:9
Birth of Kenan	3677	70	325	910	2767	Genesis 5:12
Event	Date BC from creation	Age when first son	Age of Adam at event	Age when Died	Year when died	Bible Re; Chap and verse
Birth of Mahalalel	3607	65	395	895	2712	Genesis 5:15
Birth of Jared	3542	162	460	962	2580	Genesis 5:18
Birth of Enoch	3386	65	616	365	3021	Genesis 5:21
Birth of Methuselah	3315	187	687	969	2346	Genesis 5:25
Birth of Lamech	3128	182	874	777	2351	Genesis 5:28

Death of Adam	3072				3072	
Birth of Noah	2946	498/502?	Age of Noah at Event	950	1996	Genesis 7:6
Birth of Shem	2448	100	498	600		Genesis 11:10
Methuselah died same year of the flood!	2346			969		
Year of the flood	2346		600			Genesis 7:11, to 8:13-14
Birth of Arphaxad	2344	35	602	438	1906	Genesis 11:12
Birth of Shlar	2309	30	637	433	1776	Genesis 11:14
Birth of Eber	2275	34	671	464	1711	Genesis 11:16
Birth of Peleg	2245	30	601	239	1906	Genesis 11:18
Birth Reu	2213	32	733	239	1874	Genesis 11:20
Birth of Serug	2183	30	763	230	1853	Genesis 11:22
Birth of Nahar	2153	29	793	148	1905	Genesis 11:24
Birth of Terah	2124	70	822	205	1819	Genesis 11:26
Birth of Abram	2054	101	Age of Abram at event	175	1779	Genesis 12:4
Death of Noah	1996		1996	950	1996	Genesis 9:28
Isaac	1953	60		180		Genesis 21:1-5
Abraham died	1879		1879	175	1879	Genesis 25:7
Esau	1893			147	1746	

Jacob	1893			147	1746	Genesis 49:33
Levi				131		Exodus 6:16
Kohath				133		Exodus 6:18
Amram				137		Exodus 6:20
Moses	1629?					

The above timespan list shows that Adam would have passed the records on to his children from the time he walked with God in the Garden of Eden, then how he did wrong, and how he and Eve were taken out of the garden to live in the wilderness, and this was passed on right up until Moses wrote the early history of modern life.

Because Adam lived for 950 years, 126 years before Noah was born, he would have passed this history onto future generations with Enoch his grandson, Kenan great-grandson, Methuselah his great-great-grandson, and Lamech his great-great-great-great-grandson. Any of these would have mulled over their history, and any of these grandsons who were alive when Noah was born would have shared the past with him. However, none of these grandchildren survived the great flood; they died either before or during the flood.

Ironically, Methuselah, who was recorded as the oldest man to live – to the ripe old age of 969 – died the year of the great flood. It makes you wonder, did he die of old age or in the flood itself? Was this a coincidence or meant to be? Then consider that Noah was still alive when Abraham was born, and died when Abraham was 58 years old. Up until then it is likely their history was passed on through word of mouth by just three people: Adam to Lamech; Lamech onto Noah; Noah onto Abraham. And at that time when they had the tools for writing, the history was passed onto Moses – all in a space of under 200 years.

It was inspired by the Holy Spirit! Exact in every aspect and exact in every detail! Otherwise, how would Moses know about Adam and Eve and all his family's genealogy? God instructed Moses to write it down as a history from the beginning of creation as they knew it – whether it was directly from Himself, or through Moses's ancestors, or both. Also adding into the equation, I'd point out that the first recorded writing was about 2,000BC, so writing must have been invented prior to this time. That means that some of the history could have been written and recorded prior to Moses, and could have been written when Noah and Abraham were alive.

The Bible was written over a period of 1,600 years. The Old Testament was written between 1,450BC and 420BC – a period of 1,030 years – and there was a space of 450 years before the New Testament was written at AD60.

1) Unique in its accuracy.
2) Unique in being our first history book of reference.
3) Uniqueness in its talk about supernatural events. Miracles: what evidence?
4) The New Testament.

(1) Uniqueness in its accuracy

Scientific Evidence

In Robert Dick Wilson's book, *A Scientific Investigation of the Old Testament*, Moody Press. He was born in 1856 and one of the foremost scholars in the history of the Old Testament Bible. Taking the trustworthiness of the Scriptures back to Old Testament times, he wrote: 'In 144 cases of transliteration from Egyptian, Assyrian, Babylonian and Moabite into Hebrew and in 40 cases of the opposite, or 184 in all, the evidence shows that for 2300 to 3900 years, "that is 2000BC to 400BC", the text of the proper names in the Hebrew Bible has been transmitted with the most minute accuracy. That the original scribes should have written them with such close conformity to correct philosophical principles is a wonderful proof of their thorough care and scholarship; further, that the Hebrew text should have been transmitted by copyist through so many centuries is a phenomenon unequalled in the history of literature.'

Robert Wilson also adds that 'there are about forty of these kings living from 2000BC to 400BC. Each appears in chronological order'. With reference to the kings of other countries, no stronger evidence for the substantial accuracy of the Old Testament records could possibly be imagined than this collection of kings. Mathematically, it is one chance in 750,000,000,000,000,000,000,000, that this accuracy is mere circumstance.

Thus, Wilson concludes, 'The proof that the copies of the original documents have been handed down with substantial correctness for more than 2,000 years cannot be denied. That the copies in existence 2,000 years ago had been in the like manner handed down from the originals is not merely possible, but, as we have shown, is rendered probable by the analogies of Babylonian documents now existing of which we have both originals and copies, thousands of years apart, and scores of papyri which shows when compared with our modern editions of the classics that only minor changes of the text have taken place in more than 2,000 years and especially by the scientific and demonstrable accuracy with which the proper spelling of the names of kings and of the numerous foreign terms embedded in the Hebrew text has been transmitted to us.'

The Dead Sea Scrolls

They not only added to the accuracy of the Old Testament Bible, but also authenticated it. The earliest written translated script we had was written about AD900, but then the Dead Sea Scrolls were found in 1947, nearly 20 years after the death of Robert Dick Wilson. These are believed to have been written in 125BC, which precedes the earliest script by 1025 years! This amazing find of some 500 books just confirmed the correctness of the Bible.

For instance, in the Book of Isaiah, Chapter 53, there are 166 words and the only difference was one additional word – the word 'light' in verse 11 – when compared with the book written in 900 AD. And the additional word made little difference to the meaning of the verse.

Some of the above paragraphs have been taken from Josh McDowell's book." Evidence that demands a verdict".

2) Unique in being our first history book of reference

Is it a history book? Is it the first history book ever written? There are so many history books, but none like the Bible. However, many people reject it as a book of history and just look at as a collection of fictional stories. But when you look at it in depth, it is so much more. Much of our everyday life comes from the Bible – our seven days in a week; 24 hours a day; our laws based on the ten commandments. It is also a guidebook of how to live your life, for the Jews, the Christians, and the Muslims.

There is no other history book that has had more impact on our civilisation than the Bible. I do believe that as we reject the Bible and seek to write our own bible to fit into what we want in our lives, our Sundays are not special any more, our sexual morality has changed, our belief in marriage is not the same. Our government has changed it by changing our laws. God made man and woman, and the words we use in the marriage vows state it is a sacred institution between a man and woman, not man and man or woman and woman.

There is no longer the freedom to teach Christianity in schools. What are people so afraid of? The Bible is a history book, isn't it?

Two hundred and fifty years ago, the Bible was considered our main guidance and reference book. If this is not the history book of the Jews, Christians, or Muslims, then what is it? And where would that leave these faiths – are they meaningless and just fantasy? And why were so many wars and battles fought in the name of the God of these beliefs? I'm not trying to justify these wars, though many have tried to use the Bible as justification for their actions.

If we didn't have the story of Jesus, then we wouldn't have many of our Christian holidays, including Christmas and Easter. OK, we might have pagan holidays instead, but these Christian holidays bring much joy into our lives, are something children look forward to and remember into old age. We would not have Christian schools and hospitals, and even now much of our humanitarian and charity work is inspired or directed by Christian organisations. There is so much more one could say about what came from the teaching of the Bible that one could write many books about. In fact, there are many thousands of books already written about it, including part of this one.

How many paintings have been inspired by the Bible? How many wonderful churches, and amazing worship songs?

No matter how much people try to write the Bible out of history – and in many cases have tried to burn it out of existence – it will always survive as the greatest history book ever! It's the oldest history book we have, the most read and the most printed book year after year, and published in more languages than any other. What an accolade!

Then there are the prophecies in the Bible still happening in our modern times. Take the time just after the First World War in 1948, and a strange series of events. It was agreed that the Jewish nation would be allowed to reclaim Israel as their home, their own country. This decision was a treaty made between Britain and Russia, and something the Jews had been fighting for as the rightful heirs to the land for centuries. The Jewish nations' hearts were set on this, ever since they had been dispersed to the four corners of the earth.

This promise comes from the prophecy in the Book of Zechariah (Chapter 2, v6 to the end): 'The Lord says, "Come away! Flee from Babylon in the land in the north, for I have scattered you to the four winds. Come away people of Zion, you who are exiled into Babylon!"

'After a period of glory, the Lord of Heaven's Armies sent me against the nations who plundered you. For I said, "Anyone who harms you harms my most precious possession. I will raise my fist to crush them and their own slaves will plunder them." Then you will know that the Lord of the Heaven's Armies has sent me.

'The Lord says, "Shout and rejoice, O beautiful Jerusalem, for I am coming to live among you.

Many nations will join themselves to the Lord on that day, and they too will be my people, I will live among you, and you will know the Lord of the Heaven's Armies sent me to you. The land of Judah will be the Lord's special possession in the Holy Land and He will once again choose Jerusalem to be his own city. Be silent before the Lord, all humanity, for He is springing into action from His holy dwelling.'

Was this prophecy fulfilled just by chance, or was God using this event after Hitler had tried to eradicate the Jews out of existence? Or did God somehow guide the British and Russian leaders – an unusual union – what to do? Or was it logically and expediently the best thing to do at that time? OK, there were large amounts of displaced Jews at that time, but would it not have been easier to just integrate them back into their own societies?

3) Supernatural events in the Bible: what evidence?

There are many miraculous events recorded in the Bible that cannot be 'humanly' explained, but I just want to concentrate on two of the most dramatic events in our human history. Both have profound evidence that they actually happened: the Great Flood; and the Plagues of Egypt.

The Great Flood, in Genesis, Chapter 6

Because man's sin so angered God, He flooded the world to destroy all land-living creatures.

Many geologists generally agree that there was a worldwide flood about 4,500 years ago. It is also strange that geologists found that most of all trees did not survive 5,000 years ago

Who was saved in the flood? There are accounts of families from different continents: an Aboriginal family in Australia; a family from Masai in East Africa, where a father led his family to a safe place; a Hawaiian family being saved from a great flood; and a family survived in North America. All these and the Noah story stem from the same period of time.

Were all these legends invented to justify how their families survived a great flood and helped to populate the world as we know it today? It's strange how all their stories date back to 4,500 to 5,000 years ago. Some people have suggested that the early missionaries influenced their stories by telling them about the Noah's Ark story. But their legends existed before they were evangelised. Unless they were part of ancestors of Noah's family? After all, they must have escaped the flood, too.

If just two or three of these stories are genuine, that would have given credence to a worldwide flood, especially as they cover every continent. Even if none of the above stories are true, there is still a wealth of archaeological and historical evidence of a mighty flood in Asia, around the Mesopotamia area, again four-and-a-half thousand years ago.

Noah would have had no knowledge of what was happening in other continents of the world, otherwise he would have passed that information onto Abraham and it would be recorded in the Bible. In addition, the Bible only records the history of the Jews.

If families were saved from the flood in China, Australia, Africa, Hawaii, and North America, this could explain why the world's populations survived and expanded.

The above is just a theory based on sketchy evidence, but it would explain how our world was repopulated 5,000 years ago, if we are prepared to consider that other events did happen outside the limits as set out in the Bible. Does not God love all people everywhere?

Putting the above historical and geological studies into the equation, does this not give us a strong case for the account of a worldwide flood in the Bible as being a real event, and put it beyond reasonable doubt?

There are many archaeological surveys written about different ice ages that have occurred over the last 100,000 years, often contradicting one another. But the majority agree that there was a huge flood around 4,500 years ago, in every continent of the world. What caused it, though, is a mystery.

There was a belief by some geologists that because humans were now using crops like wheat and also rearing cattle as food, this caused the carbon dioxide to increase in the air so much that it caused the icecaps to melt 10,000 years ago; ultimately causing a huge flood.

I have a problem with this theory, because geologists claim that our last ice age was for a period between 21,000 years and 11,000 years ago. So, the population of the world would have been fairly low at that time, or even non-existent. Even if the world was populated and people were burning fossil fuels, its impact would have been very low. Certainly, if you consider what we are doing in our world in this day and age, it is highly unlikely that what happened then would have been enough to even start melting the icecaps. Remember, they were just coming out of an ice age, so the sea's waters were slowly getting back to normal not vastly increasing.

Many other geologists and scientists have offered theories – some based on facts – about the reasons for the sudden huge rise in water levels. But the majority say something dramatically happened about 6,000 years ago 'by chance'.

Just around the same time, Adam and Eve were starting a new civilisation, and then 4,500 years ago, was at the time of the Great Flood! But no-one can give a reason why within a space of one year the waters rose thousands of feet, then receded in the same year. If it was the melting of the icecaps, that would have taken thousands of years! If it was a tsunami triggered by an earthquake or meteorite, that would be over in a matter of days, and unlikely to be worldwide! Yet it just happened in the year Noah completed building the ark? How amazing! Or just coincidence?

Another phenomenon was the first recorded appearance of rainbows! This was mentioned in the Bible as a promise to the Jewish people that He (God) would never flood the world in the same way ever again. The appearance of rainbows was also reported for the first time in other parts of the world where people were said to have been saved from the great flood.

There is no natural reason for such a flood, with a sudden influx of water. So, where did it come from? And where did it go? That leaves us with the possibility that it must have been a supernatural happening!!

With all the above evidence, it still needed an immense act of faith to believe in the many miracles in the Bible. Surely, it takes faith, if not madness, to build the then biggest boat ever, on a hill, miles away from water? That's also, of course, as long as you believe the Bible story and it was not invented just to fit the historical event. Or was it invented to convince people that God is real, and of the evil of their sin?

Does it take faith to believe what the Bible says? Or does it not take even more faith *not* to believe in what the Bible says?

Many would have asked why God would want to destroy so much of what He created. He loved them so much. He gave them free will! They rebelled!

In many countries, the most serious crime is that a man commit treason against their king and country, with the ultimate penalty of execution and death. In this case, man rebelled and committed treason against the God who had created them, and He is the Supreme King over everything! If it became so evil and rebelled against him, should He not have the right to destroy it, and cleanse what He had made? What if we commit treason against man? How much worse is it if we commit treason against God?

Why did God reduce the age of the lifespan of people from the high hundreds (Adam, 950 years old) down to 120 years. (Genesis 6, v3.)

The age of man was reduced to 120 years because of man's sin. That would also account for God's anger when He destroyed much of the world's population.

The Plagues of Egypt and exodus out of Egypt

The Bible is recorded as having been written between 1450-1410BC, and talks about the ten plagues, the exodus of the Jewish nation out of Egypt, and the birth of that great nation and their independence. It is believed that 70-75 of Jacob's relatives were led into Egypt, but in a space about 430 years there were believed to be 600,000 men and their wives and children. That means there must been a nation of well over one million people, maybe as many as two million people, who left Egypt and went through the Red Sea into their promised land.

In 1613BC, a massive volcano was recorded – the largest eruption in the history of modern man – and known as the so-called Minoan eruption at Santorini, a small group of islands in southern Greece, which was at that time one of the most influential and wealthy parts of the then known world. This is believed to have happened in or around the same time as the plagues of Egypt and could account for most, if not all, of the different plagues that could well have blighted Egypt. It is one of the most studied volcanoes and considered the most mysterious eruptions of all time.

Since 1969 there have been many intense archaeological excavations carried out, some beneath a layer of tens of metres of thick white ash which covers most of the islands and completely changed their outline. Below the ash they have found the Minoan town of Santorini, buried for over 3,500 years.

This volcano would have had devastating effects in many parts of the world; some of the oldest known trees to be found on earth are in California, and their year rings are said to be narrowed for a few years around the same time as the volcano. The effects of a vast cloud of dust could have blackened the surface of the earth with terrifying consequences, and its effects could have accounted for the plagues of Egypt. In the Nile river turning to blood red have been due to the dust cloud fallout? could this in turn have caused the frogs to abandon the river? this in turn could have caused the population of gnats to explode. and evented the flies to multiply feeding off the dead frogs? Boils could have been from using water from the Nile And so on. All of the plagues could be related Santorini Volcano. I'm not saying this is the reason for the Plagues, but it *could* have been the reason, unless the Biblical story was just made up to fit into the circumstances brought on by the volcano. Or could God have caused the volcano at that time to bring judgement on Pharaoh, and used Moses in bringing the Plagues to happen just at the right time? Only God knows what really happened. But whatever happened, I believe this was a real event. If the last Plague, of the death of the first-born child in every house in Egypt, never happened, it would have changed our modern life. The Jewish people were told to kill a lamb and daub its blood over the doors of their houses so that the angel of death would pass over their house. They were told to celebrate this event every year, and this is what they do; it is still known as the Passover meal.

This is why Christians celebrate Easter always at the same time as the Jewish festival of Passover, and it is on a different date every year. Jesus was called the Passover lamb, as He shed his blood on the cross at the same time as the Jewish festival. God saved the Israelites from the Egyptians, and Jesus died to save us from our sins on the cross. If these events never happened, then millions of people have been living a lie for thousands of years – both Jews and Christians. Easter would not exist, and our children would have no Easter eggs!

When Moses led the Israelites out of Egypt, away from slavery into the desert, he knew that somehow God would find a way for them to cross the Red Sea and to enter the Promised Land. If that event never happened, the state of Israel would not have existed today. Or was this whole event just a fairy tale, so to speak?

God parted the Red Sea. The Red Sea was over seven miles wide and in many places half a mile deep. Nothing but a miracle could have saved them.

God uses water in miraculous ways

There are many events recorded in the Bible where water was used in miraculous ways throughout history. In the Old Testament Bible, I have already talked about the great flood with Noah, where most, if not all, of the world was suddenly covered with water and then vanished in a short space of time. Then the waters in the Nile turned to blood red. And as part of the exit of the Israelites from Egypt, the waters of the Red Sea – being half a mile deep and seven miles across – suddenly opened up, allowing over one million Israelites to cross to the other side. Then, when they were all safe, the waters suddenly closed, trapping the whole of the Egyptian army, which was swallowed up and drowned. And again, if we look at when the Israelites were in the desert for seven years after the exodus from Egypt, God split a rock in two and water gushed from it.

In the New Testament there are also many accounts of where Jesus used water in miraculous ways. The first miraculous event using water was when Jesus was baptised in water and the Holy Spirit fell on Him.

The first miracle He performed was at a wedding in Cana in Galilee. Jesus and His mother Mary were at the wedding and she told Him that they were running low on wine. This is quite shameful in the Jewish custom, so Jesus told them to fill six stoneware pots to the top with water and fill them to the brim. The water in each pot, which holds twenty to thirty gallons, was changed into the finest of wines. They said about this, 'Who keeps the best wine to the end of the festivities? Rather one serves the best wine first and the worst till last, when the guests were worse for wear.' This was the first of Jesus' miracles. There are many other occasions that Jesus used water in the miracles He performed, including walking on water, stilling a storm at sea, and others.

I had an experience when I was a young Christian and we were members of a small country church just outside Maidstone in Kent, back in the early 1980s. We had monthly renewal meetings when like-minded Christians from different churches in the area got together. Our church, like many other country churches, had no direct water supply. It was our turn to host the monthly meeting and, naturally, to supply hospitality to the guests, serving teas and coffees in the mid-interval. We filled the water urn up some time before meeting to allow the water to boil, using water from the vicarage a few hundred yards up the road. When we started to serve the teas and coffees, there were a lot more people than normal, so we ran out of hot water with only about half the people served. We looked inside the urn and it was nearly empty, so I put the lid back on and said a little prayer, knowing it would take ages to get more water from up at the vicarage then wait for it to boil.

We carried on pouring hot water till everyone had been served. At the end of the meeting, I looked inside the urn, and to my amazement it was still about quarter full. OK, it did not turn into wine, but how did this happen? And why?

On another occasion, our Christian Youth Club went for a day's outing to the coast. We had this day earmarked to spend time on the beach at Camber Sands, near Rye in Kent. The day came and the weather forecast was for rain across the whole country. Should we cancel? We decided to go and say a small prayer.

It was raining the whole way down there, but as soon as we arrived at our destination the rain stopped. It turned out to be a beautiful day – one of those days that was not too hot, but just perfect, with a slight haze and not even a breeze. It stayed dry the whole day and we even had a BBQ. It was as if the day had been made just for us and others at Camber.

Some of our group decided to go into the nearby town of Rye for a few hours, but as soon as they went down the road a mile or so, it started to rain and didn't stop until they nearly arrived back.

It started to rain again on our way home, after such a wonderful day. If Jesus can calm a storm and control the elements 2,000 years ago, why not today?

I'm not saying in my personal life that the above events are anything to write home about, but it just proved to me that some things can't be explained, humanly speaking.

There are so many other events in my life that can't be explained, and for which science has no answers. From a broken ankle made whole, money arriving at just the right time to meet a need (sometimes in excess of £1,000), and one time when our youth group was given miraculous powers to ward off a marauding gang of skinheads – two of whom were carted off to hospital! But that's another story.

There are a multitude of miracles that have happened all over the world, and which cannot be explained. In a church in Timor, Indonesia, back in the 1980s, they experienced an amazing outpouring of the Holy Spirit. It was in a small village, where it seemed like fire was coming from the top of the church. It was a very poor community, and on another occasion during Holy Communion they filled a cup with water and blessed it. When they took the Communion, somehow it had turned into wine. They experienced many other miracles from walking on water to eating deadly poison with no effect.

Sometimes, I believe God just wants to show us how real He really is, and just wants to bless us.

Two of the most dynamic events that I've mentioned in the Old Testament were the Flood, and the Plagues of Egypt. Another amazing event is the missing day in the world's history. With all three, throughout the earth and in all the different continents, there have been recorded catastrophic events that have mirrored the events in the Old Testament Bible – also in the same periods of time! Many of these could be just myths, ancient legends, or folk tales, but it is strange that they have all come from different sources, and many can be confirmed by scientific and geographical evidence.

God's Power Given to Defeat Armies, and the Missing Day in the World's Calendar!

Before we talk about this amazing event of the missing day, we must set the scene of what it was like, the belief that was given to Moses and his people at that period of time, and for the Israelites to take hold of the land that God had promised them. This was after they had miraculously escaped from the captivity of the Egyptians and had been wandering in the desert in Jordan for forty years.

Towards the end of Moses' life, when he was 120 years old, he went up to the top of Mount Nebo, which overlooked Jericho and extended from there to the Mediterranean Sea. The Lord told Moses this was the land He promised under oath to give to Abraham, Isaac, and Jacob. Moses had laid his hands on Joshua, and he was full of the Spirit and wisdom, so the Israelites followed him and accepted him as their new leader.

After Moses had died, the Lord spoke to Joshua and showed him all the land the Israelites were going to inherit, with a promise that no-one would be able to stop them from taking the land.

To enter into the Promised Land, they first had to cross the River Jordan. Now, we're talking of in excess of two million people, so the Lord told Joshua to select twelve strong priests – one from each tribe of the Israelites – to carry the Ark of the Covenant. The Ark was 45inches long and 27inches wide, made of acacia wood, covered with pure gold, and housed the two tablets on which God had inscribed the terms of the Covenant between God and His people. It had four gold rings, which were for two poles to be inserted. No-one was allowed to touch the Ark, as it was so holy; if they did, they would instantly die!

They took the Ark to the edge of the Jordan, and as soon as the first priest's foot touched the water, it backed up until there was a clear path for the priests to carry the Ark into the middle of the river on dry land. When all the Israelites had crossed the river and the Ark was safely across the other side, the waters began to flow again.

If you had experienced this happening twice in your life, in the space of just forty years, first, the crossing of the Red Sea and now the crossing of the River Jordan, surely anything was now possible?

God then defeated Jericho when their walls miraculously came tumbling down.

One of the next battles the Israelites faced was against the Amorites. As they were retreating, and being pursued, the Lord destroyed the Amorites with a terrible hailstorm which killed more of them than the Israelis killed with the sword.

Joshua, Chapter 10, v12-14.

'On the day the Lord gave the Israelites victory over the Amorites, Joshua prayed to the Lord in front of all the people of Israel. He said, "Let the sun stand still over Gibeon, and the moon over the valley of Aijalon."

'So the sun stood still and the moon stayed in place until the nation of Israel had defeated its enemies.'

Is this event not recorded in the Book of Jashar? The sun stayed in the middle of the sky, and it did not set as on a normal day. There has never been a day like this one before or since, when the Lord answered such a prayer. Surely the Lord fought for Israel that day!

Not only was the mighty Egyptian army defeated by God drowning them in the Red Sea, but while the Israelites were in the desert for forty years, they had built a strong army who were quite small in number, compared to others in the region and sometimes outnumbered ten to one. Yet they defeated every army that lived in the promised land. A total of thirty-one kings and their towns were destroyed, often in miraculous ways, until the whole region was claimed! If God is for us, who can stand against us?

I have experienced this miraculous power that is my life. On one occasion when I was a fairly young Christian, we had a Church Youth Group who often saw God working miracles. My eldest daughter Kim's sixteenth birthday was coming up and, being in the seventies, she was a Mod. She wanted a special party with all her Christian friends, so I booked a fairly small hall in a village just outside Maidstone, near where we lived. On the day of the party some of her friends were in the town, dressed in their parkas, when a gang of skinheads had decided to meet in the town from the surrounding area. There was a slight altercation between them, but not wanting any trouble, my daughter and friends made themselves scarce.

Somehow the skinheads got wind of the party. And after all my daughter's friends had arrived and the party got underway, one of the group reported that the skinheads were gathering on the other side of the field, opposite the hall. Not wanting my daughter's party to be spoiled, I went into the pub next to the hall and phoned the police, informing them of the situation. After about fifteen minutes, several police arrived and warned the skinheads then left. But this warning was not heeded, and as soon as the police had gone, they amassed massed a stone's throw away from the hall.

We had already prayed for protection for the party. But they were armed with bricks, bottles, cans, sticks, and stones, and as we came out to see what was happening, they started to shower us. I felt a calming peace and authority, and shouted, 'Go in the name of Jesus!' They stopped in their tracks, turned round, and retreated.

We hoped that would be the last of the matter, but no. One of the party guests, who had been invited but was not one of the youth group, went out and started to taunt the skinheads, so much so that they decided to have another go. Our amazing youth just went out in God's power and completely overpowered them, without a single scratch being laid upon our group. I went over to their ringleader and grabbed hold of him. It was like holding a baby chick in my hand; I had complete power over him. That night, at least two of them were taken to hospital!

I'm not saying this was anything like Joshua's battles. God has never spoken to me face-to-face like He did with Moses and Joshua, but to me it was a real demonstration of God's power.

However, God did audibly speak to me on one occasion when I was walking through Sloane Square in London, and changed my whole course of life. As a result, I gave up my directorship in a London-based company to work with the homeless. We bought a large house and had up to eight young men – aged between fifteen and twenty-one – come and live with us. This was a venture of faith, and we were totally reliant on God to supply our needs. No church, no charity, just God! But that story is for another time.

The Missing Day!

This is either the most outrageous deception or the most amazing miracle in the Bible, apart from creation itself. How could it happen, unless you had the power to create the world and create time. This could only happen if there is a force that created and controlled the Universe! If you don't believe in the Bible and God, then this would just be a made-up, supernatural story.

The great flood with Noah, and the Red Sea parting to allow the Israelites to escape from the Egyptians, are miracles confined to the limits and laws on earth. But stopping the world in its tracks is so mind-boggling it is impossible to comprehend, unless you had power and authority over the stars and galaxies themselves.

Let's go back to the question: what is a miracle? If it is something that defies logic and science, unless it's some form of magical deception, then it's outside our human understanding, is of a spiritual nature, and confined to either good or bad spiritual forces. It might be just a small miracle, like finding a needle in the haystack, so to speak, or a broken bone suddenly being healed. Or in this case, the day standing still for twenty-four hours! But they are all outside our human understanding, and one is no more miraculous than the other. Miracles are miracles, it just depends on your faith! Some are big, some are small.

There are many theories on how the missing day happened and, if it did, what effect it would have on the world if it was worldwide. Was it just being written poetically to the earth rotating slower, or stopping altogether? Some of the theories may have an element of truth in them, but when it comes down to it, you either believe it is one of God's miracles or you don't. Some things we are just not meant to understand. It is only a fool who can think he can understand God and His powerful works!

One can argue and debate till you are blue in the face, but you will not get an answer, unless you could speak to God Himself!

There are records of parallel events to be found in other countries in different parts of the world. Ancient Chinese writings speak of a legend of a long day. The Incas of Peru and the Aztecs of Mexico have similar records. Babylonian and Persian legends tell of a day being miraculously extended.

Herodotus, an ancient historian, recounts that while in Egypt a priest showed him a temple record of a day which was twice the natural length of any other day!

The dates of these records don't marry up, but they are all in the same millennia and, so far, I have not heard of any period where a long day has been recorded. This in some ways to me only adds to it being an actual historical event.

Chapter Seven

The New Testament. Who is Jesus? Is He a real person? And how Christianity has changed the world

There are literally thousands of books written about Jesus and the New Testament Bible, and I could write a book on each chapter, but that is not the object of this book. It is to establish the facts that Jesus was and is a real person. He is the Son of God, and the New Testament is a true account of His life while He walked on this earth, how in the life of the Disciples of Jesus He set out to establish a new understanding about God, in Christianity, how it radically changed the world, what impact it had on society 2,000 years ago, and how it shapes our history even to this day.

Let's presume for one minute that the New testament Bible was not a true account of the person of Jesus Christ, but a series of stories invented by a group of four individuals. They called themselves the disciples of Jesus Christ – Matthew, Mark, Luke and John; one a tax collector, and two others were fishermen. Luke was the only non-Jew (a gentile) and he was a physician with a profound knowledge of the Greek language. They were an unlikely bunch, as they all came from different walks of life, considering they were both labourers and intellectuals. One day, they all met up in the local tavern, and I can't imagine how the conversation started as they somehow set out to invent the most elaborate hoax ever, and write a book which would deceive the whole world into thinking the main character in the book. Jesus

Christ, was the hero who was coming to save the whole world. He was a Superman! Not conceived naturally, but sent from God – a cosmic force from another world. I can imagine them talking about different topics in the Torah, which is the Old Testament Bible; and one old hat topic being who and what and when the new Messiah would come. So, they decided to make the story come to life.

This book was so good that they all put their names to it as the authors. Luke became a close friend to a man called Paul, so he asked Paul – a learned Jew, but also a freeborn Roman citizen – to write the second part of the book. He would tell of how he was hoodwinked into this new cult by having an amazing conversion. But more about that later.

After discussing their basic story, they went away to invent what part each would play in this intriguing plot, and what kind of friends they were to our superhero. Then they invented eight other friends to play other parts of the story. They also had to be either trusted friends or close relations, so all would be friends and followers of Jesus – the Magnificent Twelve. They also chose one of them who would be an imposter, a cheat, an informer who would ultimately betray Jesus, leading up to His death. His name would be Judas!

They all got together around the year 50AD and set about writing this most amazing book ever, which they completed in around ten years. They met in absolute secrecy, for they knew that if news got out of what they were doing, their lives would be in severe danger. They were taking parts of the Holy Book and using them for their own means, which was punishable by death. They were also proclaiming Jesus to be the King, so if Herod got wind of this, they could have been arrested and executed for treason.

In a space of 200 years, this new religion had thousands of followers from all over Europe, and parts of the Middle East,

Asia, and Africa. Then in the Year 337AD, Emperor Constantine of Rome was converted to this new belief in Jesus. The Roman Empire was vast, so from then onwards many believed in this new religion and its new leader named Jesus, who had left them by dramatically ascending to a place called heaven – this 'other world' – to be with God. When He left, He sent the Holy Spirit, who was able to be everywhere at any time, though invisible. And if you swore allegiance to Jesus, you had the ability to invite Him to live in you and use His power to change the world.

Ultimately, this new religion spread to all corners of the world, with its followers being prepared to die in their plight to spread the word. And many did.

I'm digressing again!

Back to the story. We'll make this story quite short, so this superman Jesus would only live to the age of 33. We won't say much about the first 30 years of his life either, except for the story of his birth and where He came from.

Our book will be full of intrigue, the most amazing love story, a story of passion, hate and violence, a horror story of how He was to be strung up on a cross and left to die, for all to see. Hope for the future; He must have been the greatest trickster, miracle worker or illusionist. Have power over death; power to defy gravity, to walk on water, power over the elements to calm a storm, to heal the sick, to feed thousands from just two loaves and four fishes, to turn water into the finest of wines. And a promise; if you believe in Him, you can do the same tricks as Him, and even better tricks by using the power of the Holy Spirit, who is now living within you. Also, when your earthly body dies, your spirit and soul will live with Him forever… IF… you follow Him! If not, you will be judged when you die and your soul could go to a place called hell. Ultimately, He will destroy all evil in an apocalyptic happening when He returns for the last time, and everybody will see and believe.

There will no book to match it ever. But hey! This book was written 2,000 years ago. Not even Shakespeare can even remotely compare with it.

Back to the story again. Sorry, I keep digressing.

We must dispel any idea that this Jesus was a madman, going around just pretending to be a New Messiah. This Jesus is the Real McCoy, so to speak, with completely new ideas about God and faith. There were many other new ideas going around at that time. Even the Muslims talked about a man called Jesus performing miracles; a good man, and a prophet. They've recorded Him in their bible called the Koran. So, the story must conform to what they say about Him, even if they didn't acknowledge him as the Son of God.

We must also dispel any thought that He was a liar, a cheat, and imposter, trying to deceive people into thinking He was something special just to get money and to get them to worship Him. What was He doing it for? We must not show that at any time He ever asked anything for Himself.

One minute. One of the prophecies said He went into Jerusalem riding on a donkey! We must invent a fantasy story of how He asked one of us to go into town and find the donkey tied up outside the village tavern, and if you are asked what are you doing stealing my donkey, just say, 'My master needs it!' So, on that occasion He did ask for something, but that was to fulfil one of the prophecies.

We will make up other stories of how He did ask people to bring Him something, but it will always be to assist Him to perform a trick or a miracle. Like any superman, they are always the champion of the people, never doing anything for his own benefit and ego. After all, He was going to pay for it all in the end by giving up His life, supposedly for us all.

He – that is Jesus – must fit the picture perfectly. We'll base the story of Him being a Jew, so the story of His life must conform not only to our Hebrew teachings in the Old

Testament but also to more than 300 prophecies written about Him in the Torah, about the coming of a new Messiah, a new king who will save them and deliver them, setting them free from their tyrannical Roman rulers. We must be careful not to leave any of their prophetical teachings unturned; otherwise the story will be seen as unauthentic.

He was born a Jew, from the line of David. Right, Matthew. You start by writing a brief history of His genealogy through to Mary, OK? His family tree line comes from Abraham, so put in all the genealogy up to David. Then put in that there were fourteen generations from David up to Mary; fourteen being a special number for the Israelites. Because Mary was a young Jewish lady, they only go by the male's heritage, but let's assume she was of the lineage of David and Abraham. That would be a good excuse not to write all the names of her last fourteen descendants. People will just accept that the prophecy is correct and that Mary was a direct descendant of Abraham and David. 'That will fulfil some of the prophecies in Isaiah,' it says in Chap 7, v14. The Lord himself then will give you a sign. Look! The virgin will conceive a child! She will give birth to a son and will call him Immanuel, which means 'God is with us'. Also in Chap 11, v1 there's mention that a stump of David's family will grow a new shoot – yes, a new branch bearing fruit from the old root.

Then in Isaiah 9, v6-7: 'For a child is born to us, a son is given to us.

The government will rest on His shoulders.

And He will be called: Wonderful Counsellor, Mighty God. Everlasting Father, Prince of Peace.

His Government and its peace will never end.

He will rule with fairness and justice from the throne of His ancestor David for all eternity.

The passionate commitment of the Lord of Heaven's Armies will make this happen.'

Jesus' character must conform to the above prophecies of our superhero. Quite a tall order! Not even Abraham, Moses, David, or Joshua could in any way meet these qualifications.

Though Joseph was not the biological father of Jesus, we will write how he was betrothed to Mary and was told by an angel that it was OK to marry her as God was the father of Jesus.

Luke agreed to write a complete list of Jesus' male ancestors, as Jesus was known to be the son of Joseph, even though he was not the biological father. Starting with Jesus and going right back to Adam, the first son that God created, right up to Jesus – the second son that God created in a supernatural way.

Mary lived in Nazareth and Jesus was to be born in Bethlehem in a stable; Mary was a virgin when she conceived Jesus by the Holy Spirit. That would cover five more of the prophecies.

We have to put in that He came from Bethlehem, because there is a prophecy in Micah, Chap 5, v2: 'But you, O Bethlehem Ephrathah, are only a small village among all the people of Judah.

Yet a ruler of Israel will come from you, one whose origins are from distant past.

The people of Israel will be abandoned to their enemies until the woman in labour gives birth.

Then at last at last his fellow countrymen will return from exile to their own land.

And he will stand to lead his flock with the Lord's strength, in the majesty of the name of the Lord His God.'

Good! What next?

Right. There is another prophecy about in the Torah in Isaiah 43, also in the last chapter in the Bible, Malachi 4, about someone coming, preparing the way for the Lord.

Let's build a story around this messenger. Let's call him John, and explain how he was born in quite an amazing way, about the same time as Jesus' birth.

So Luke created this intriguing chapter partly about John's life, by telling how and when he was born, what a peculiar kind of person he was in how he lived his life, how people looked up to him and were fascinated by this new teaching, leading into the coming of the new Messiah and who He – that is the Messiah – would be, and what people should do in preparation of when He comes; the way John the Baptist ultimately dies at the hands of the Roman rulers, by being executed in a most bizarre and horrific way and his head being served up on a platter because of the evil plot of Herod's wife. Herod was at that time the Roman ruler over Galilee. It was also agreed that Mark would collaborate with Luke and write his version of John's death at the hands of Herod.

The story of John's parents goes like this: they were a very elderly couple named Zechariah and Elizabeth, never had children, and were now well past the age of conceiving. Zechariah was a God-fearing man, and one day while he was in the temple an angel visited him and told him not to be afraid and that his wife Elizabeth would give birth to a son, they should call John. The angel said the child would bring Zechariah great joy and be filled with the Holy Spirit even from birth, but he must never touch wine. He would be a man of Spirit and power; the power of Elijah – a revered and holy prophet in the Torah.

John was very blessed by being filled with the Spirit of God, and lived in the wilderness and fed on a diet of just locust and honey until he was in his twenties; then he went around telling people how to find salvation through forgiveness of their sins. The only other way at that time was by sacrificing animals, so this was a completely new way. From then, they were to be baptized in water, to show they had repented and had turned to God to be forgiven. John was to warn people wherever he went how to live good lives if they wanted true salvation; this would be a completely new way of living, by sharing their food and clothes with the poor. Many were puzzled, as at that time they were expecting the new Messiah to come soon, and would ask John, 'Are you the one we have been waiting for?' But he answered their question by saying, 'I baptize you with water; but someone is coming who is far greater than I am, I am not even worthy to be his slave let alone even tie the straps of his sandals. He will baptize with the Holy Spirit and fire.'

One day, when John was baptizing crowds of people down by the river, Jesus joined in. But when He was baptized, the heavens opened and the Holy Spirit, in bodily form, descended in the form of a dove, with a voice from heaven saying: 'You are my dearly loved Son and you bring me great joy.'

That will fulfil some of the other prophecies in Isaiah 11, v2-5:

'And the Spirit of Lord will rest on Him; the Spirit of wisdom and understanding, the Spirit of council and might, the Spirit of knowledge and the fear of the Lord.

He will delight in obeying the Lord.

He will not judge by appearance, nor make a decision based on hearsay.

He will give justice to the poor and make fair decisions for the exploited.

The earth will shake at the force of His word, and one breath from His mouth will destroy the wicked.

He will wear righteousness like a belt and truth like an undergarment.'

I sometimes forget whether I am writing this as a real event that actually happened, or a story invented by the disciples?

Back to the story:

As we have created our superhero Jesus, we must also talk about the other force that's always with us and was told in the first chapter of the Torah: a force of evil, and how they were at war with one another. These two heavenly forces are going to do battle and use us as their battleground, fighting for our souls: one to guide us into making good decisions, leading us in a way that God created us for; the other continually tempting us to do bad, wicked things. But because this force of evil, whom we will call the Devil, wants to lead us away from believing in God and Jesus, he will try to get us to believe that they don't even exist. He will try to deceive you, as he did with Eve in Genesis by tempting her to do what God specifically told her not to do. Before then, everything was rosy in the garden, so to speak, because both Adam and Eve had no understanding of what wrong was.

This new part of the story starts when Jesus was 30 years old. The battle begins. After Jesus was baptized, the Holy Spirit would lead Him into the wilderness for forty days, where He would face His first battles with the Devil. No matter what the Devil tempted Jesus with, quoting different parts of the Old Testament Bible, Jesus replied with other Scriptures that overrode the Devil's temptations. From then on, the Devil knew he had no hold on Jesus and that Jesus was invincible, though he tried and tried.

Then finally, he thought he had triumphed when he induced Pilot to kill Jesus by hanging Him on the cross – only to find Jesus had even defeated death, by rising from death three days later. He had been slaughtered, but this was the greatest happening ever. Our superhero proved that even a physical death could not hold Him. He would be the greatest man to ever live, and by dying for us He had taken our sins on his shoulders and left them in hell.

Earlier in the story, while Jesus was with them, He told the disciples how to receive eternal life. (John, Chapter 3, verse 16) 'For God loved the world so much that He gave His one and only Son, so that everyone who believes in Him will not perish but have eternal life.'

And as the story goes, they were not to fully understand what He told them until after He had died and risen from the grave.

We must build the climax by creating a story that is the time leading up to Jesus' death and resurrection, as there are many more prophecies said about Him, as recorded in the books of Psalms, Isaiah, Zechariah, and others. We'll create this story built around those prophecies.

He was despised and rejected by His own people. (Isaiah 53, v3) He was despised and rejected – a man of sorrows, acquainted with deepest grief.

We turned our backs on Him and looked the other way.

He was despised and we didn't care.

He was dumb before His accusers. (Isaiah 53, v7) He was oppressed and treated harshly, yet he never said a word.

He was wounded and bruised for our sins. (Isaiah 53, v5) 'But He was pierced for our rebellion, crushed for our sins.'

He was whipped and spat upon. (Isaiah 50, v6) 'I offered my back to those who beat me and my cheeks to those who pulled out my beard. I did not hide my face from mockery and spitting.'

He was mocked. (Psalms 22, v7-8) 'Everyone who sees me mocks me. They sneer and shake their heads, saying, "Is this the one who relies on the Lord? Then let the Lord save Him! If the Lord loves Him so much, let the Lord rescue him!"'

They pierced His hands and feet. (Psalms 22, v16) 'My enemies surround me like a pack of dogs; an evil gang closes in on me.

They have pierced my hands and feet.'

He was crucified with thieves. (Isaiah 53, v12) 'Because he poured out Himself to death, and was numbered with transgressors.'

He prayed for His persecutors. (Isaiah 53, v12) 'He bore the sins of many and interceded for rebels.'

'They divided my garments and cast lots.' (Psalms 22, v18) 'They divided my garments among themselves, and cast lots for my clothing.'

'They offer me gall and vinegar.' (Psalms 69, v21) 'They offer me gall for my food and sour wine for my thirst.'

'Why have You forsaken me?' (Psalms 22, v1) 'My God, my God, why have you abandoned me?

Why are you so far away when I groan for help?'

They pierced His side. (Zechariah 12, v10) 'Then I will pour out my Spirit of grace and prayer on the family of David and on the people of Jerusalem. They will look on me whom they have pierced and mourn for Him.'

Darkness over the land at noon. (Amos 8, v9) '"In that day," says the sovereign Lord, "I will make the sun go down at noon and darken the earth while it is still day."'

Was put in a rich man's tomb. (Isaiah 53, v9) 'He had done no wrong and had never deceived anyone.

But He was buried like a criminal; He was put in a rich man's grave.'

Then, three days later, we will concoct one of the most amazing events in history – how Jesus was dead and had somehow become alive! He had actually risen from being dead! He was alive! And many people would witness this fantastic event. They had seen Him horrifically die on the cross, yet three days later He was walking and talking to them. Though, somehow His appearance had changed, but He showed them his hands and feet to prove He was Jesus their Messiah; He ate with them and told them what to do when He had gone to be with God in heaven.

What did our superhero say about sin? What could be better than the ten commandments? Let us cut it down to two. What could replace them? Love your neighbour as you love yourself, and the other, Love the Lord your God with all your heart, your strength, soul and mind. If you do this, that will cover all the old commandments! Who would steal from himself?

After Jesus ascended into heaven, Luke wrote this amazing story, about their adventures of being filled with the Holy Spirit and how they went out into the streets telling people of what had happened to them, that they, too, should follow in the footsteps of Jesus and be baptized. They prayed for the sick and they were healed, in the same way that Jesus had shown them. Luke also wrote about his friend Paul and his unbelievable conversion, leading to the start of Paul's historical journeys throughout Asia and Europe in the next chapter in the story. He called this next chapter The Acts of the Apostles.

A large part of the book of Acts will be centred on Paul's missionary journeys and will be recorded in great detail. Luke will accompany Paul, as his companion, and that will account for the story's accuracy.

On Paul's last journey to Rome, while he was being transported by ship under the guard of a Roman commander, Luke changes the narrative from being centred on Paul to 'we' in Acts, Chapter27, implying that he travelled with Paul. This carried on throughout the chapter and would account for the great detail of their perilous journey, right down to the exact record of 276 people on board a boat that was transporting them to Rome via Malta. If you were inventing a story, why would you suddenly change the narrative to include yourself? Was this an autobiography of Paul's life in Acts, from his dramatic conversion into becoming a follower of the Christian faith, or just one big fairy tale?

As the story goes, written by Luke, Paul believed this new cult were heretics, 'being a Jew himself'. His name at that time was Saul, but after his conversion he changed it to Paul. On one occasion, he witnessed and agreed to the stoning of Stephen – 'one who was selected by the disciples to be an apostle, a teacher'.

Saul went to the High Priest to obtain permission to arrest and kill the followers of Jesus. So, the story goes (in Acts 9, v5) that Saul was on the road to Damascus when a bright light from the sky suddenly shone down from heaven. He fell to the ground and heard a voice saying, 'Saul! Saul! Why are you persecuting me?'

'Who are you, Lord? Saul asked.

And a voice replied, 'I am Jesus! The one you are persecuting. Now get up and go into the city, and you will be told what you must do.' The men with Saul heard someone's voice but saw no-one! Saul got up from the ground only to find that when he opened his eyes, he was blind!

His companions led him into Damascus. After three days, a man come to see him – a believer called Ananias, whom God had told to go to Saul. God forewarned Saul that a man named Ananias was coming to lay hands on him so he could see again.

Ananias had heard rumours how Saul had done terrible things to believers, but God assured him that Saul was his chosen instrument to take His message to the Gentiles and to Kings, as well as to the people of Israel. 'I will show him how much he must suffer for My name's sake.'

So that is what happened. When Ananias laid hands on Paul, he instantly saw, and something like scales fell from his eyes. He could see again, but he was also filled with the Holy Spirit. Then he got up and went to be baptized and changed his name from Saul to Paul. He passionately dedicated the rest of his life to converting both Gentiles and Jews, with many fascinating journeys. Wherever he went, many believed, and many miracles followed them, with new churches established.

One of these journeys began when he was arrested in Jerusalem, and followed his travels to Rome and its importance. Paul was to take Christianity to the Romans in Italy, but how was he to get there? Because Paul's teachings were so radical, in so much as he was opening the door for Gentiles ' that they could become part of the Jewish family by believing in this new Messiah, this was completely unacceptable to the Jewish leaders. Their law says all men must be circumcised and abide by their rituals of scarifying animals for forgiveness of their sins.

So, Luke created one of the most intriguing court dramas ever – one which was to last for several years, and was a battle between the laws of God, and the laws of the land! The Jews made several plans to kill Paul. Their first attempt saw them drag him out of the temple but, as they were about to kill him, the Roman guards intervened and arrested Paul, bound him in chains, and took him to the fortress. They were pursued by the angry Jewish mob, shouting, 'Kill him, kill him!' The Commander allowed Paul to speak to the crowd. Paul could speak in Greek and Aramaic, and he spoke to the crowd and explained how he was a zealous Jew from Tarsus (that is in Turkey), and was trained in their laws. He also had the authority to persecute the followers of the Way, hounding them to death or throwing them into prison. He continued to tell them of his amazing conversion, how he had been blinded and Jesus spoke to him, telling him to be baptized and to go tell the Gentiles all about Him and how they could also be part of God's family. With that, the crowd were incensed, shouting that he was not worthy to live.

Then Paul revealed to the Commander that he was also a Roman citizen. When the Commander found this out, he became frightened because he had ordered one of their own to be bound and whipped. The Commander wanted to know why the Jews were so incensed, so the next day he ordered the Jewish High Council to come to a meeting with Paul to find out what they had against him. When Paul started to put his case, proclaiming his innocence, there was an altercation with Ananias, the High Priest. Paul, well versed in the Jewish law, cleverly pitted the Sadducees and the Pharisees – two factions of the Jewish council – against one another. One claimed Paul to be innocent, the other guilty. This caused such an uproar, with Paul in the middle, that the Commander ordered his soldiers to rescue Paul again and take him back to the fortress.

For several years, the Jews continued to plot to kill Paul, but he remained in the custody of the Roman guard. Each time he was brought before the court, in several places and heard by different Roman governors in different provinces, none could find any reason for Paul to be punished, let alone be beaten or killed. Eventually, after several years of imprisonment, Paul put in an appeal to Caesar, the Roman Emperor. So King Agrippa ordered that Paul was to be sent to Rome and Caesar, to be judged.

They left on board a ship with several other prisoners, all in the custody of a Roman officer and his men and the ship's crew. In total there were 276 on board. Being the winter, it was not the best time to sail, and after many weeks at sea and many storms, they were shipwrecked off the coast of Malta, where miraculously not one of them lost their lives. Paul was bitten by a venomous snake, but it did not harm him, which amazed the islanders. Then he healed many of the island's sick people, so the people on the island honoured Paul with gifts and supplied the travellers with all they needed for the rest of their journey to Rome. They spent three months on the island of Malta.

When Paul finally arrived in Rome, there were Christians who welcomed him. He was allowed his own lodgings, though he was guarded by a soldier. No action was taken against Paul, and he was allowed to continue with his mission to preach to the Romans.

After we have written all our versions of the story, we will get together to make sure they generally agree with the details in the storyline. It won't matter if there are slight differences, as that will make the book sound more authentic and show we have not just copied each other.

This is where Luke's and the disciples' epic stories end, and Roman history takes over.

There are several records in history that tell us that Rome had a great fire, on the exact date of the 18th of July, AD64, under the rule of Emperor Nero, and that Christians were established as part of the population then. It is said that Nero blamed the Christians for the fire in which seventy percent of the city was destroyed. So he tortured and executed many of them including Paul, who was beheaded. The date of Paul's death is said to be July AD64. The remains of his body are buried in the Basilica in Rome.

Other records say Nero started the fire so that he could re-build part of the city. There is the famous quote that 'Nero played the fiddle while Rome burned'. It was probably the lyre, as the fiddle wasn't invented till 1500 years later.

Two things are part of Rome's history: There was a great fire in AD68; and Christians were part of that history. Tacitus, a Roman aristocrat and historian who was a young boy at the time of the Great Fire, wrote about all the events at that time, including the Jewish sect called Christians.

Nero, at the age of 30 in AD68, is recorded as having committed suicide. I wonder why? Judas did the same!

Many places that Paul evangelised, like Athens, Malta, Cyprus, Rome, Jerusalem, and many other places, still to this day lay claim to Paul's visits to their land as an important part of their history. There are other places that choose not to remember, but they are mainly where the main religion is Islam, although in some places relics still exist of early Christian churches.

You may ask why I have placed so much importance on the person of Paul. Well, nearly half of the New Testament Bible was either written by Luke about Paul (his conversion and where he went on his evangelistic missions), or by Paul in his letters to all the countries and cities that he had visited and had established new Christian churches. These letters never contradict one another; there is no negativity, only positivity, always giving Glory to God, Jesus or the Holy Spirit, and setting out to give the new Christians teachings on how to live their lives.

If I set out to write such letters, I would have to have an ideology to base them on. I'd need knowledge of all the countries going up through Lebanon, Syria, Turkey, into Greece, and culminating in Italy, also the Mediterranean Islands of Cyprus, Crete, Malta, and many others. If it hadn't been for Paul or someone like him, Christianity could never have spread as rapidly as it did. Cause and effect!

To base all these stories on fabrication is more unbelievable than believable. Luke records all the places that Paul evangelised, and Paul's letters correspond to these places; there are no discrepancies.

Can you seriously believe that Matthew, Mark, Luke, John, Paul, and others, concocted this amazing story that not only conforms to all the prophecies in the Old Testament, changed the world two thousand years ago, and is still changing our world today with more believers than any other religion? It also has credence from the Jews, Muslims, Romans, and many others, as it is recorded as part of their histories. They showed how you could be executed for claiming to be God 'in the form of being Jesus the New Messiah', and for implementing this new ideology with Jesus. The penalty was execution on the cross. They were also promoting others to join this new club, whose members called themselves Christians. Writing such a book was an extremely dangerous venture... unless their writings were true, their experiences were real, they were disciples of Jesus, and their beliefs and faith were stronger than life itself. Unless they did actually walk, talk, eat, and live with Jesus, witness His death on the cross, His resurrection, and continued to live with Him in His resurrected body for 40 days. Unless they also witnessed His ascension into heaven, experience the coming of the Holy Spirit in the upper room that gave them boldness to go and preach the Gospel, baptising people as Jesus did, healing the sick, and performing amazing miracles. Everything that gave birth to this new religion called Christianity.

With such a weight of evidence, I might suggest your only reasons for not believing must be spiritual or intellectual, for the evidence stands as proof! Jesus spoke in parables that confused the minds of intellectuals, but children could understand with ease. These were illustrations, giving comparisons of different scenarios to show what the Kingdom of Heaven was like. Did the disciples have the wisdom to create such stories?

Here are just two smaller examples to explain what I'm talking about. The Parables of the Hidden Treasure and the Pearl. The Kingdom of Heaven is like a treasure that a man discovered hidden in a field. In his excitement, he buried it again and sold everything he owned to get enough money to buy the field. In the other, the Kingdom of Heaven is like a merchant who was on the lookout for choice pearls. When one day he found a pearl of great value, he sold everything he owned to buy it. What would you do to understand the mysteries and secrets of eternal life?

Did Matthew, Mark, Luke, and John, really invent all these parables in the Bible? If they did, surely they must have been filled with wisdom outside their own comprehension? And what about Paul? Was his conversion from being a persecutor of the Christians to probably the greatest evangelist ever, just fabrication?

Take just eight of the Jewish prophecies about the coming Messiah from the Old Testament: place of birth; the time of His birth; the way and where He was born; how He was betrayed; the way He was executed; the people's reaction; the way He was pierced; and His burial. Now, look to see if anyone in history could fulfil all of these prophecies. The possibility would have to be at least 1 in 100,000,000,000,000,000. If at all possible.

In Jeremiah 23, v5-6, under the heading, Righteous Descendant: '"For the time is coming," says the Lord, "when I will raise up a righteous descendant.

He will be a King who rules with wisdom.

He will do what is just and right throughout the land.

And this will be his name: The Lord is our Righteousness.

In that day Judah will be saved, and Israel will live in safety."'

Who else in history has ever fitted the bill? Other than Jesus!

Josephus, Flavias was a well respected Jewish historian (AD30-100). At the age of 19, he became a Pharisee; in AD66 he was a Commander of the Jewish forces in Galilee. He wrote a lot about the history of his time, including an explicit account about John the Baptist that not only he was a real person but he was going around baptizing people for purification of their bodies, 'not for the forgiveness of their sins'. He gave a detailed account of how and why John was beheaded at the hands of Herod. Although the reasons varied between Josephus and the Bible, that just adds credence to this being a real historical event and not just a made-up story.

He also wrote a hotly contested quotation, from the Jewish Rabbis: 'Now there was about this time Jesus, a wise man, if it be lawful to call him a man, for He was a doer of wonderful works, a teacher of such men as receive the truth with pleasure. He drew over to Him both many of the Jews, and many of the Gentiles. He was the Christ, and when Pilate, at the suggestion of the principle men among us, had condemned Him to the cross, those that loved Him at the first did not forsake Him; for He appeared to them alive again the third day; as the divine prophets had foretold these and ten thousand other wonderful things concerning Him. And the tribe of Christians so named from Him are not extinct at this day.' The Arabic text of this passage adds: 'He was perhaps the Messiah concerning whom the prophets accounted wonders.'

It would have been very controversial to suggest that the Messiah had come in the form of Jesus. Josephus must have been a man of considerable authority to write such events and get away with it, to the dissatisfaction and anguish of the Jewish Sanhedrin.

Is all the above just fabrication? Or, are they true events that have been handed down to form part of our rich history? Even if just 10% were true – and I am not suggesting that is the case – it would mean that the historical man, Jesus, is not just a myth. He existed as a historical figure. If so, who was He? The Son of God, or just a good man?

Who was He?

In this chapter about Jesus, I came to the conclusion that I could not talk about Jesus unless I included the Holy Spirit as well, because they both testify to each other's existence and they also testify to an all-powerful God. I will explain more about the Holy Spirit in the next chapter.

At the time 2,000 years ago when Jesus was born, the Old Testament Bible included over three hundred references or prophecies to the coming Messiah. The Jewish nation at that time was ruled by the Romans, and the Jews were waiting for God to send a new ruler to set them free from the tyrannical Roman rule. It was prophesied in the Bible that the Messiah would come as a baby and born in Bethlehem, born of a virgin and from the line of David. One of the first accounts of the genealogy of the Jewish Nation goes back to the first book of the Bible when, in Genesis 22,18, God tells Abraham that 'all your descendants and the nations of the earth shall be blessed, because you have obeyed my voice'.

This is one reason why the Jews were so fanatical about keeping their genealogy records exact. If you wanted to discredit a person, you would try to destroy all evidence that a person existed in history. To discredit Him, the easiest way would be to take Him out of their history book and any reference to Him, but this would mean taking out over three hundred prophecies that alluded to the coming Messiah. This would be totally against their laws and religion., unless they believed that He had not yet come and they were still waiting for Him, (i.e. it was not Jesus).

However, since they have now been restored back into Jerusalem, this does not conform to the timescale, so the Messiah must have already come. So, who could He have been? Have we missed Him, or was it Jesus? The Messianic Jews believe it was Jesus! There is now a church of 350,000 Jewish Messianic believers worldwide. Some 2,000 years ago, they were looking for a King, a strong and powerful ruler, and though the Bible warned them that they would not recognise Him, somehow they failed to see Him for who he was.

Morison, a barrister and engineer, was an atheist who wrote a book called *Who Moved the Stone*, published in 1987. I first read this book soon after I became a Christian; it deepened my faith, and confirmed my belief that Jesus was real.

The book showed a long chain of evidence that Jesus was a real person and is the Son of God. Morrison's exploration of early historical records proved to be an excellent starting point of spiritual exploration.

The strangeness of the Resurrection story was so extraordinary in its conception that it captured his imagination and why sceptics were so interested that a single person could actually change the future history of mankind. He set out to prove that this extraordinary story of Christ's Resurrection was just a myth, and if it was a myth, who could have written and made up such a story. He decided that if Jesus was a real person and if by chance He did exist, then Morrison could prove He was just an ordinary person.

The author set about gathering all the evidence, but the more he gathered, he found overwhelming evidence that not only did Jesus exist as a true historical person, but He was who He said He was, the Son of God. This led him not only to believe in Jesus but also that God Himself is real.

His findings led him to discover the validity of the biblical records. *Who Moved the Stone* is considered by many to be a classic on the subject of the Resurrection of Jesus Christ; it includes a vivid and poignant account of Christ's betrayal, trial, and death on the cross, as a precursor to His amazing and climatic Resurrection. It also looks at all the events around the Resurrection, what life was like at that time in history, including the laws between the Romans and the Jews, and the intriguing events that unfolded leading up to and after the Resurrection. This was either one of the most elaborate hoaxes or the greatest true story ever.

As a lawyer, he has written a well-researched book, fascinating in its detail to reason and accurate accounts leading up to and following Jesus' Resurrection, showing it to have been an actual historical event that never was a myth.

If this New Testament story book about a person called Jesus was created by a group of people some 2000 years ago, why is it still the bestseller ever? It is still the most read book on every day of the year. Why do billions of people all over the world sill base their beliefs in the person of Jesus? The Bible is not filed under Fiction, but considered a true story of the life of a real person who lived and died and overcame death, with no grave to mark His death, with nothing for people to grieve over, because He is still alive in people's hearts. It is also an intricate part of our cultural history.

Who could invent such an elaborate story, based on the current history of those days and more than three hundred prophecies from the previous 1,500 years of prophets to testify that He was real? Why is our yearly timescale based on AD and BC, and all our main holidays based around the biblical festival where the main character is Jesus? Why have so many people been prepared to lay down their lives for their belief in Jesus? Was this all in vain, and for nothing? Why are Christians the most persecuted people in the world? Why were our first schools and hospitals started by Christians, because of their belief in Jesus and his teachings? Why is our culture based on Jesus' teachings of Love your Neighbour? Why have we built our most beautiful buildings, churches, and cathedrals, just to worship Jesus? Why do Christian charities still give far more in aid to the poor than any other charities in the world? Why is the cross the most worn emblem in the world, and how many have it incorporated in their national flag? Why are the names of God and Jesus Christ the most blasphemed names? What is so different about these names? No-one says 'O, Buddha!' or 'O, Mohamed'. If you did take Mohamed's name in vain, you could have your head chopped off!

When a country follows the ways of Christianity, it prospers. I fear for our country in the way it is going! It is a battle for good and evil, so where does 'good' come from? Just take out an 'O' from the word 'good'!

Our world is becoming more and more greedy. At the time of writing (March 2020), we are at the beginning of the Corona virus pandemic. People are so fearful that they are just stripping the shelves of the supermarkets, with no consideration for others. The Christian way is to give the poor, and not to think you are any better than another.

OK, I know that Christians have acted wrongly in the past; being a Christian does not make you perfect, but it does make you aware of the many wrongs not only in your life, but also in others' lives. The Christian way of life is to say sorry and mean it. There are many people who do things in the name of Jesus or Christianity, but which are sometimes for their own benefit. Being a Christian and following Jesus is the most perfect way of life. It not only blesses you and your family in this life, but gives you a hope for and in eternity.

Is all of Christianity built on a basis of made-up stories to deceive billions of people, or has it got a supernatural power, beyond our understanding, that is in charge and driving us forward? If it is all fabrication, it has done an amazing job!

What is God's plan for the next generations to come and, ultimately, the end of this world as we know it, leading to a new world far beyond our imagination. You can read some of what it is going to be like in the last book of the New Testament Bible, Revelations.

Chapter Eight

Is there a force for evil?

Let's look at what is evil? Before the knowledge of Adam and Eve some six thousand years ago (that's if you accept the concept and existence of Adam and Eve), the first thoughts were recorded of the existence of a force of evil, Knowing right from wrong. If there is a force of evil, it logically follows there must be an opposite – a force for good. Do these just exist, or have they been created? Would there be knowledge of good and bad in prehistoric life? I don't think so. Both primitive man and all living creatures lived and based their existence on the survival of the fittest, and the law of nature was not based on reason.

If we have a creator God and you look at what has been created, it is beautiful and perfect. Unless there is a creator, evil has no meaning. But if this is so, why would God create evil? Or was there another force, not of His doing? But then, He is the creator of everything, so why and what is behind this force that we call evil? I'm afraid that is one question I don't think we can ever fully answer.

However, if there is a force behind bad, what form does it take? The Bible talks of a personal devil. How can we evaluate this? If there is a force for evil, then it stands to reason that there is a force of good, and as these two forces are opposites, they must be against one another -yes? A possibility, or probability, is that these two forces were in existence together before time as we know it, then if the Creator created human beings (man) in His own image, which is 'perfect', then this other force could have been very jealous and sought a way to infect God's pride and joy! This evil set out to find a way to infect man's mind with information, with knowledge of evil. Until then, man would only have known good.

There are three elements in modern man: the primal animal instinct; the force of good; and the force of evil. Compare this to just an animal instinct force, which was the law limited to prehistoric man and animals.

Adam and Eve at first had no other knowledge of anything good or bad, only what their Father had told them. So when the Devil came along, in whichever form you care to believe, their minds must still have been pure. They'd had absolutely no outside influence, never met anything else of a spiritual force, so in other words had no way of knowing what evil was, though they were told by their Father not to eat from the tree of knowledge. How would they know if the Devil was good or bad in what he was telling them? Bad was not a concept! Would I have been tempted? After all, God imparted a creative mind to this new modern man. He would not have even mentioned it to Adam and Eve unless He knew they would be tempted, and this was part of His overall plan for the future world. If this temptation had never happened, this world would either be a place of perfection with no knowledge of evil, or it would be a place just like the prehistoric world with no understanding of God!

Would you have denied Jesus three times to save your life? As the disciple Peter did. When Adam and Eve did wrong and met in the garden with God, they instantly knew they had done wrong and knew they were naked. When the greatest purity comes up against evil or wrong, it's like chalk and cheese. A child knows when it has done wrong when Mum or Dad comes into the room, as opposed to any other living creature.

Chapter Nine

Mind, Body, and Spirit

The Mind

The mind is so complicated. It is the hub of everything we are, the driving force behind our existence, the 'computer' that controls all we do and think. How would you go about creating such an ingenious tool? Science has come a long way, but it still does not know how the brain works, other than that it sends the correct signals to each and every organ in the body, how to move our limbs, even the nerve endings that tell us what is damaged in the body. What would it be like if you could not feel pain?

It controls our creative mind, using our imagination in writing and painting, and how to make amazing things from the invention of the wheel, building bridges, medical science, to fathoming equations, building computers, and so much more. The list is endless.

It also controls our emotions: how to love, hate, cry, show pity and compassion, to empathise, and so much more.

In the last fifty years we have created the computer, and tiny chips with immense data stored in them, but what or who created the mind?

One of the main research studies in modern day science is how to repair brain disorders like dementia, epilepsy, and other neurological conditions. My wife has Alzheimer's. But as long as she can think, love, can control her movements and have emotions, she is just as much a person as you or me. And even if she loses these abilities, her spirit is kept safe in God her creator. That is not to say that I don't stop praying for her healing.

We could write a whole book about the mind; it truly is the most amazing organ in our body. We can live without a limb, or sight, or hearing. We can even repair the heart. But not the brain! What or who gives me the words to write this book?

The Body

Carrying on from the previous passage on the mind, the skull encases the brain and protects it from harm.

We have learnt how to make things mechanically with metal and plastic, but not with tissue, blood, and bone. We have created robots and taught them to think for themselves and be creative, but they do not possess feelings, such as being hurt either physically or emotionally. Earlier in the book, I mentioned that we are learning how our DNA works and how to genetically modify its structure. But we've hardly started figuring out how to create it from scratch.

Our bodies are perfect in every way, and work in perfect harmony with our mind unless we have a defect in our being which we call illness.

Can you think of any other function that could possibly improve our lives, other than the five main senses we have – smell, sight, touch, hearing, and taste? We also have two other senses, called the vestibular and proprioception senses. These are to do with balance and where our bodies are in relation to space, the ability to clap our hands, walk a straight line with our eyes closed, to know when we are standing up or lying down, or what pressure or force to use to open a door, to throw a ball and judge its length to reach its intended destination.

Some people believe that we are created as part of Natural Selection, that somehow our bodies evolve to adapt to the situation or need that it finds the best possible way. I personally do not understand this way of thinking. Maybe I'm missing something. Where did all this Natural Selection and the intelligence from a tiny cell come from?

There are many books on the body written by biologists, scientists, and others, which point out what amazing creations we are. Needless to say, I'm not qualified in any way to write more about the wonderful make-up of our bodies.

Spirit and Soul.

There are many elements that God has used in His power to create this wonderful world of ours. It is difficult to understand whether it is God the Father God, Jesus the Son of God, or the Holy Spirit that has created our world and everything in it, and in the heavens above, which all work in complete harmony together. For example, if I pray to God through Jesus in the power of the Holy Spirit, and a miracle happens – who made the miracle work? They are three in one and one in three. Our prayers are known by all three!

Let's get back to the different elements of spirit. God created the heavens and the earth and everything in it by the power of His Spirit. In the first Bible chapter, Genesis, the Spirit of God hovered over the surface of the water.

When we look at every living creature, each has a soul and that soul has a spirit as part of its make-up. Some might argue that only humans have a spirit. But if we look at animals, we know some have emotions which display the spirit they possess in their being. For instance, elephants are known to weep, swans mourn for their lost mate for life and never find another partner, many animals play, and some protect their young unto death.

Dolphins express joy and have been known to protect others. They have even been known to save people from danger and enjoy playing with them; they somehow have an inbuilt relationship with us that connects their spirit with our spirit.

Years ago, when I was on holiday in Lanzarote with my wife Jane and my daughter Sarah, we decided one day to go out on a catamaran to see the dolphins. We were having a great time and were well on the way back, but had so far seen no dolphins! Sarah and I were dangling our feet over the side of the catamaran and discussing why we had not seen any, and I suggested we should pray about it. As soon as I said, 'Lord, show us some dolphins', they instantly appeared under our feet! I know dolphins have this amazing inbuilt sonar system, but is there a connection between their spirit and ours? Or was the above just a set of coincidences?

All animals make decisions, to go left or right, to choose which partner to mate with, and they all have a spirit and mind to be able to make their own individual decisions. They are not robots; each has its own DNA.

So, if we work on the concept that every creature has a soul and a spirit as part of its make-up, it therefore follows that we are not just mind and body but we have this other element that cannot be seen or scientifically explained called the spirit. But what part of the spirit do we have that all other creatures don't have? Well, we have to go back and look at the previous chapters. In particular, Chapter Five explains what happened 6,000 years ago with Adam and Eve, when God breathed His creative Spirit into man. From then onwards, man had this creative spirit built into his DNA; this supernatural element that far exceeded anything that prehistoric life had, or had ever possessed. This element of the spirit was split into two. The first was when God walked with Adam and Eve in the garden and there was no barrier between them, no sin. His spirit was pure. They talked to one another as a father would talk to his children, but as soon as they ate the apple, which was from the tree that bore the knowledge of good and evil (whether this was an actual fruit-bearing tree or a way that God created, like a parable to help us understand the concept that too much knowledge can be good or bad) that put an enmity between man and God. From then onwards, man had a choice: he could choose between a life dedicated to God with all its benefits; or a life feeding his own ego, desires, and greed. His spirit became tainted. This also opened the door for the force of evil to affect or even control his minds and actions. People say someone has a good or bad heart; in essence, the heart and soul are the same. Man still had this creative part to his spirit, but he lost that intimate part which gave him that closeness with God.

The Holy Spirit in the Old Testament

The Jewish nation ushered this new era into the world's history. From then onwards, there were men and women who chose to either follow God and follow His precepts, or their own desires. The Bible is full of actual events that showed that when they followed His ways there was a supernatural element to whatever they did and the spirit inside them took on a super, supernatural side to whatever they did; they were filled with the Holy Spirit. The Bible is full of these super, supernatural events.

Individual men throughout the Old Testament Bible did amazing miracles in God's name. They were filled with the Holy Spirit that overtook their whole being. It was shown in many ways, like when Noah was told and given the wisdom and ability to build a boat on dry land, when

Moses was told by God to build the Ark of the Covenant to house the tablets that God had given them on which was written the laws they must abide by. In the Bible (Exodus, Chapter 31), it is written that God chose two men, Bezalel and Oholiab, and other craftsmen, and filled them with His Spirit. They were gifted craftsmen in gold, silver, and bronze, and mounting gem stones, also in carving wood. They were gifted men humanly speaking, but God chose to give them supernatural gifts in their enablement to help build the ark.

Michelangelo, who was said to be a devout person, believed that the path to God can be found not exclusively through the Church, but through direct communication with God. Looking at his work, his paintings, drawings, and sculptures, what he achieved in his life is simply staggering. Was he filled with God's supernatural gifts?

There are many other examples in the Old Testament Bible where God chose many people from the Israelites to do amazing things. They were all men and women who were devout in their belief in God, from Adam to Abraham and the prophets, and those who wrote the Bible. They were filled with God's Holy Spirit and chosen by Him. The Jewish Rabbis would not have accepted anything written in the Bible unless they believed they were proven holy men of God. Every part of the Bible has that super, supernatural element to it.

The Holy Spirit in the New Testament

All people will be filled with the Supernatural Spirit of God from now on, if they believe in Jesus and He is the Son of God.

Joel, Chapter 3, v28-29. An Old Testament prophesy:

The promise of the Holy Spirit.
Then after doing all these things.
I will pour out My Spirit upon all people.

Your sons and daughters will prophesy.
Your old men will dream dreams, and your young men will see visions.
In those days I will pour out My Spirit even on servants – men and women alike.

This started with both Jesus and John the Baptist being filled with the Holy Spirit even before their birth.

John, Chapter 14, v15-17. Jesus' Promise of the Holy Spirit:
If you love me, obey my commandments. And I will ask the Father, and He will give you another Advocate, Who will never leave you. He is the Holy Spirit, who leads into all truth. The world cannot receive Him, because it is not looking for Him and doesn't recognise Him. But you know Him, because He lives with you now and later will be in you.

John, Chapter 16, v7-9. The work of the Holy Spirit:
Jesus said, 'But in fact, it is best for you if I go away, because if I don't the Advocate "The Holy Spirit", won't come. But if I do go away then I will send Him to you. And when He comes, He will convict the world of its sin, and of God's righteousness, and of the coming judgement. The world's sin that it refuses to believe in me…'

Chapter 16, v12-15:
There is so much I want to tell you, but you can't bear it now. When the Spirit of truth comes, He will guide you into all truth. He will not speak on His own but tell you of what He has heard. He will tell you about the future. He will bring me glory by telling you whatever He receives from me. All that belongs to the Father is mine; this is why I said the Spirit will tell you whatever He receives from me.

When? Luke, Chapter 24, v49:

Jesus said, 'And I will send the Holy Spirit, just as my Father promised. But stay in the city until the Holy Spirit comes and fills you with power from heaven.

Acts, 2, v1-4. The Holy Spirit comes:

On the day of Pentecost all the believers were meeting together in one place. Suddenly, there was a sound from heaven like the roaring of a mighty windstorm, and it filled the house where they were sitting. Then, what looked like flames or tongues of fire appeared and settled on each of them. And everyone present was filled with the Holy Spirit and began speaking in other languages, as the Holy Spirit gave them this ability.

Three aspects of God's Spirit:

1) We have already established that all creatures have a spirit as part of their soul – things that science has no answer for, but which we call natural, part of nature. But in essence, everything that is living is supernatural; in fact, this is what this whole book is all about. If we believe that we have a supernatural God who is beyond our understanding, it stands to reason that everything that He created must be supernatural.

    So, the first part of God's Spirit was evident in God's creation. Every living creature has a Soul and a Spirit.

2) Adam and Eve were given a new creative part to their soul! God gave us a part of His Spirit.

    He breathed His Spirit into man. This creative Spirit was part of His nature. That is why He always knows what we do and think. God is Omnipotent: He sees our comings and goings; He knows what we are thinking, whether we have good thoughts or bad; He knows when we try to put bad thoughts or actions out of our minds; He also knows when we thank Him for even the small things in life; He is in all places at all times. This creative part of God's Spirit has been entrusted to us, and has the freedom to do good or bad. The spirit in man also has the ability to seek that God part in his soul, that part that prays to an invisible God.

If we choose to ignore God, He leaves us to our own devices, but if we choose to turn back to Him or find Him for the first time, He welcomes us with His arms open wide, like a long lost son or daughter.

So, the second part of God's Spirit is what was entrusted to this new man in Adam.

3) Then there is the Holy Spirit reserved for believers in Jesus and the ancient prophets. As I have mentioned earlier, God said in Joel, 'He will pour out His Spirit on all people'.

What form does the Holy Spirit take that was so different from anything else? First, the Holy Spirit will only dwell in those people whose lives are made pure. Adam and Eve's lives were the first, they were given the first language to speak in. That language was entrusted to the Jews but was taken away from any tribe of people that turned to their own ways and denied God. The tower of Babel. I'll speak about that and speaking in tongues a little later in this chapter. Next, the Holy Spirit filled different people who He chose, who He knew would put Him first and be prepared to lay down their lives and carry out His plans for the future world. God knew right from the start of modern humanity that He was going to send Jesus to purify the souls of people, to make them pure enough to receive the Holy Spirit. And this is precisely what He did two thousand years ago, for you, me, and the rest of the world – the seal on our lives that would guarantee our souls would be restored to Himself for eternity. How precious is your soul? Is it stamped 'sealed' with God's Holy Spirit in Jesus?

Speaking in Tongues!
Acts, Chapter 2, v5-12:

At that time there were devout Jews from every nation living in Jerusalem. When they heard the loud noise, everyone came running, and they were bewildered to hear their own languages being spoken by the believers.

They were completely amazed. 'How can this be?' they exclaimed. 'These people are all from Galilee, yet we hear them speaking in our own native languages! Here we are – Parthians, Medes, Elamites, people from Mesopotamia , Judea, Cappadocia, Pontus, the province of Asia, Phrygia, Pamphylia, Egypt, and the areas of Libya around Cyrene, visitors from Rome(both Jews and converts to Judaism), Cretans, and Arabs. And we all hear these people speaking in their own languages about the wonderful things God has done.'

They stood there amazed and perplexed. 'What can this mean?' they asked each other.

The above verses tell of just one way in which people could speak in tongues.

Speaking in Tongues takes on several different forms. First, it was the ability to speak in known languages that you have never been taught; next, there is the ability to speak in unknown languages – known as speaking in heavenly languages. The easiest way to describe this phenomenon is to let your soul be open to God so that He can speak to your spirit within you. It is that part that is the heavenly part of your being, when you accepted Jesus into your life. When you speak to God in tongues, it is your soul communicating with God; your soul speaks to your mind and brings God's words into human understanding.

I don't know how this really works, but what I do know is that it gives me words or messages in situations beyond my own wisdom. I have spoken tongues quietly and the whole atmosphere has changed, even when I have been in a public place and someone has come up to me and started to talk about God. It is that God-part in me that is seeking and wants to know more. Other times, when I don't know how to pray in a situation, and I pray in tongues, something happens. It could be that someone needs to be healed, or it could be for finance or another need. On other occasions, when there is a conflicting or you could say an evil atmosphere, I've spoken against it quietly in tongues. Then the evil physically leaves the room, and peace descends.

On another occasion, a group of business friends were using tarot cards. Again, I quietly prayed in tongues and told the one who was dealing them that they would not work, he got more and more frustrated as they would not work and asked me to leave the room.

Praying in tongues does not involve your mind, just your soul and the spirit inside of you.

Another way of describing it is that God wants to communicate with us and speak to that part in us, that heavenly part 'our' soul that was made immortal when we accepted Jesus in our life.

My wife Jane has Alzheimer's, and sometimes when she is feeling down and confused, I pray in tongues and her whole demeanour changes; a peace floods over her, her smile comes back, and so does her equilibrium. It's something I could not do with my own strength. I continue to pray one day she will be completely healed.

Another way speaking in tongues is used could be in a group of Christians, when somebody speaks out in a heavenly language and someone else translates them into a known language. This is often used in prophecy for something in the present or for the future. It could be a reprimand or an encouragement.

There is so much more that I could write about speaking in tongues and the Holy Spirit, but that is for another time.

The Holy Spirit very much at work in this day and age

I don't believe that we should limit God to what is written in the Bible. God loves everyone! Everywhere, He knows their suffering. Was He ignoring the rest of the world while He just concentrated on the Jews and that part of the world? Or were there many other incidents that happened that are not written in the Bible that God was doing in other places but are not written down? Even in the last verse in John's gospel, it is said that Jesus did so many other things (Chap 21, v25). If they were all written down, I suppose the whole world could not contain the books that would be written.

If this is written about Jesus in His three-year ministry, how much more would God be working throughout the whole world? How big is your God? There are many instances in recent years where God has shown who He was in different parts of the world where there was no Christian influence, from children in China, a Buddhist monk in Burma, a Muslim in Iraq, and many others. But there is one thing they all have in common. They all had visitations from God, and all pointed to Jesus as the Son of God! God has poured out His Spirit on all these people! His super, supernatural Spirit is only given to those who accept Jesus as the Son of God! No Buddhist, no Muslim, no Hindu, no Atheist or Agnostic, and now not even any Jew, can receive this super, supernatural Spirit of God unless he believes and accepts Jesus as God's Son. 'God sent His only Son into the world to save it and cleanse it from sin.'

Should we limit God to only what is written in the Bible? I don't think so! Does not God love people everywhere in the world? Does He not feel the grief and suffering of all of His creation?

Did He save people in other parts of world, in Canada, South America, Australia, and other parts of the world from the Noah's Ark flood?

Chapter Ten

Believing in the Natural/Supernatural

Is there a pure world based on science and physics where every question can be answered? Or is there what we call a supernatural force that created everything we know? Or could it be a combination of both? Maybe if there is a supernatural force, and if we understood more about it, it would no longer be known to us as supernatural but as the norm? As 'natural'?

That also raises another question. If everything was and is created by a supernatural being, then it follows that even a blade of grass or grain of sand is supernatural. Or the norm.

Putting the above aside, let us define what is understood to be supernatural as opposed to natural. For example, levitation defies the laws of gravity. Healing of the mind or body, where a part of the body is completely healed without the intervention of medicines or surgery – perhaps a cancer that miraculously disappears, or a broken bone that has been instantly repaired. Or it could be a material item that suddenly appears to meet a specific need or situation, a specific sum of money mysteriously is posted through the letterbox to meet a specific need, or provisions that arrive on the doorstep just at the right time, or a situation where a person missed catching a plane or other vehicle that was involved in a fatal accident. Many would class all the above examples as 'supernatural', others would describe them as fate, chance, or pure coincidences to be somehow explained away. But so many happenings seem to occur when people have a need and cry out in prayer to a supernatural God, who they cannot see or hear. How many of us, in a time of desperate need, have cried, 'Help me, God!' Or 'Save me, God!' and have lived to tell the tale?

In the Bible there are numerous happenings that tell of miraculous incidents – physical, material, and spiritual. Is the Bible just a story book with made-up folk tales? Or is it a book that speaks of a supernatural God who still wants to be involved in our lives today?

There are other elements in our world that we would term as spiritual which involve the supernatural. They defy the natural laws of our world and cannot be explained logically – some considered good, some evil, some used in different cults and religions.

What does it take more faith to believe in? That this world just happened by chance and that everything came into being from a tiny cell or cells? Where did they come from? All the elements of our world just came together at exactly the right time to sustain them?

Or to believe in a designer that created and planned all we know? Someone gave time, space, substance to this planet, bringing it into being and in complete harmony with one another?

Whatever we believe, logic tells us there must be something out there that brought our world as we know it into being, whether that was as described in the beginning of the Bible, some other force that governs life as we know it, or what other people believe as Natural Selection with no external 'supernatural' intervention. Did we just happen? Or evolve?

Chapter Eleven

Life in another dimension

If there is no life outside our world as we know it, this book is meaningless. In fact, everything is meaningless. Life outside our dimension must be a spiritual one, as if and when our physical lives die, there is nothing. It also means we are only mind and body. There must be more!

The Bible is full of spiritual input, starting in Genesis, Chap 1, v1, when God's Spirit hovered over the waters of the earth. In fact, every chapter in the Bible has a spiritual element in it, painting a beautiful picture of a spiritual God. Most other religions believe in a spiritual world by recognising that all humans have a spirit.

Many people say that man has got to have something to believe in, otherwise life has no meaning, and believing is just a crutch for life. If it is a crutch, then it's an amazing crutch, which not only gives us a hope for the future, but helps us in every part of our daily living. What a crutch!

I hope I have given plenty of evidence to prove beyond reasonable doubt that there is a God who, not only created us, but also cares for us from the day we are conceived in our mother's womb to the day our spirit departs from this world and beyond.

There must be some truth in the belief of a personal God. If not, the Jewish race did not come into being from Adam and Eve, the Muslims also did not come from Adam so have no basis for their family tree back to Abraham and Adam, and Christians would also have no foundation and Jesus would be just a fairy-tale. So, if the Jews, the Muslims, and

Christians were all deceived, it would have been the greatest hoax ever and our history over the last 2,000 years would have been totally different and, dare I say, much for the worse without our belief.

There have been many atrocities carried out in the name of religion, and many more are still happening today, but the good far outweighs the bad. I could talk about many of the wonderful happenings both in our social lives, from schools, inventions, hospitals, leadership, and so much more.

What is this world in another dimension?? This other world can only be imagined in our minds, or we can get more ideas from the Bible, if it was inspired by the Creator. I believe this to be true.

There are records of many people who have visited a world in another dimension – 'heaven' – and come back to share their experiences. I have not personally met anybody who has experienced this, but there are several books on people who have had this out-of-body experience. Some actually died and came back to life to share their profound experiences.

There must be life in another dimension, otherwise what hope is there? And this book and everything in life would be meaningless!

Chapter Twelve

My final thought for us to comprehend is: What is History?

This book not only brings our history into focus but also gives us a logical reason for our existence. We had to have a cognitive understanding of time, a way of knowing what our past was and what our future could hold, and a way of recording it. I believe before 6,000 years ago prehistoric man did not have the ability to record his history, neither could he only rely on his memory to retain events, as these memories would fade with each generation. In addition to that, I don't believe they had an audible language to converse and comprehend what history is. Was their existence the same as the animal kingdom, based on survival of the fittest, and the primitive primaeval law of the jungle, so to speak? Neither were they equipped in any way with writing materials or had the skills to record their history, even if they did have cognitive ability. If they'd had the ability to write then surely we would have found some evidence? As far as I know, the only records we have are of cave drawings of animals. No written words!

Is our history based purely on science?

If we take out of this book all the history of the Jews, including Adam and Eve and all the Christian and spiritual input, and just concentrate on the bare scientific, historical facts and geographical evidence, then the creation of the world would be full of mysteries that just could not logically be explained. For instance, how the first living cells came into existence and multiplied! The first animals, all plant and tree life, the multitude of different forces that keep our world in balance.

Geographical surveys map out the history of our world events, taking us through all the different species of animal life and all the different ice ages through millions of years. Then, just 6,000 years ago, 'prehistoric' life came to an end, and modern man went from a primitive life of living in caves to building sophisticated palaces in under 1,500 years. So, you would have to consider all the above just happened by chance in a sequence of events, or you could look to see how or if the spiritual side of events, with a supernatural creator at the centre, fitted into the equation and if the Bible gave a true record of our early historical past. To me, there are too many questions that cannot be answered without involving an exterior force, outside of pure science.

Does history only have a meaning with modern man? Or was there another force outside of modern man that had a cognitive understanding of history?

Does time and history exist?

Because we are all just a part in passing time, and presuming this world's civilisation that came into existence some millions of years ago will one day come to an end – either by a natural catastrophe, by our own doing, or by an Armageddon – then one day we will be no more. Thus, it stands to reason, time and history will not exist! Are time and history themselves not the invention of mankind? Animals, including chimpanzees, have no understanding of time and history.

So then, does it matter if we live or die? Does it matter if we do good, or bad? Evil, just like good, would have no meaning, and in fact, everything we do would be completely pointless. **Unless...** We are part of a creation, designed by a supernatural being, who has a plan for our future existence, be it temporal or eternal, whatever eternal may be. Were time and history created? Did we create them? Or did God? Did God fill us with part of Himself by planting His Spirit into our being so we could comprehend both the past and the future. If so, I believe He must want us to be part of that future and an ongoing part of history, be it in another dimension. Heaven?

Time and history are God! He is the beginning of all history and He will be the end of all history! The future is in His hands! Believe it or not! The Alpha and Omega!

Invitation

If you decide that you are convinced that God is real, and that Jesus is the Son of God and He died for you to cleanse you from all sin, then I invite you to admit that you have sinned and done wrong and ask Jesus to come into your life. Finally, invite the Holy Spirit to come and live in you and be with you, thanking our Lord Jesus for giving you eternal life.

When you have done this, please go and seek out other Christians and share what you have done, so that they can testify to what you have done and can support you in prayer that any doubt may not enter in.

I would also recommend that you join a Christian church and enter into an Alpha course, if they do one.

Interpretation

This book is written for each reader to interpret however he wishes. You can throw it out as a discarded garment. You can take it as a mental exercise, or an aid to the knowledge you already have. Or you can use it to understand more about the spirit inside you and the wonderful world we live in.

I hope many will take it on board and seek a closer relationship with our creator, and that it will lead you to know God in a new dimension.

Whichever way you use and read it, I pray it will bless you and open your eyes, heart, and mind to the wonderful world we have been given to live in.

Many people live in fear of the future and have no idea of where we come from and what the future holds.

This book is written to give people a hope for the future and a purpose for their existence. It also gives an insight into the history and creation of our world from concept, through the different times and stages of its formation. The book is not written in complex language, so almost anyone can understand it. It also sets out good arguments that can confound the most learned of our society.

I've used information from scientists, historians, geologists, and philosophers, that can challenge some beliefs, myths, and theories, and exploring whether we evolved from tiny cells or is there a creator of our Universe.

So, whatever you believe, ask yourself: Does it logically make sense? And if we say to ourselves, I just don't know or I'm just not sure, then this book will give you a much better understanding to work from. It could set you on a life-changing journey. I'm not saying it will give you all the facts, as there are many things in this world we just do not know, but it will give you basic facts of the world's history and what an exciting, beautiful, and wonderful world we live in.

Index

Abel 49
Abraham 42, 46, 47, 56–7, 59, 66, 84, 100
Abram 58
Acts of the Apostles 91
Adam 48–52, 105
  and the Bible 56–9
  creation of 31, 36, 42–4, 46, 47, 54, 57
  death of 58
Agassiz, Louis 25
Agrippa (Herod Agrippa I), King of Judea 94
Amorites 75
Amos, Book of 90
Amram 59
Ananias 92, 94
apes 13, 14, 21–2, 29
Ark of the Covenant 74–5, 111
Arphaxad 58
Assyrian Empire 60
Athens, Greece 95
Attenborough, David 11–12
Aztecs 78

Babel, Tower of 114
Babylonia 60, 61, 64, 78
Bethlehem 85
Bezalel 111

Bible, the 25, 28, 30, 37, 38
  accuracy of 60–62
  credibility of 41, 55–60
  as historical record 62–4, 65
  language of 42
  New Testament 60, 71, 80, 96, 111–13
    see also Jesus Christ
  Old Testament 44, 60–62, 83–84, 96, 99–100, 110–11
    see also Genesis, Book of
Big Bang Theory 1–2, 3, 8, 24, 28
Brodgar, Ness of 31

Caesar, Julius 94
Cain 49
Cambrian Explosion 22, 25–6, 28
cellular life 11–12, 15, 18, 33, 35
chimpanzees 13, 14, 21–2, 29
China, ancient 31, 78
Christianity 25, 42, 63, 70, 97, 121
chromosomes 14
climate change 6–7
Collins, Francis, *The Science of God* 8

Constantine, Emperor of Rome 82
Coronavirus Pandemic 103
Cox, Brian 24
Creationism 2, 3, 8–9, 10–11, 28
Cyprus 95, 96

Damascus 92
'Dark Matter' 2
Darwin, Charles, *Origin of the Species* 10, 13, 14
    Darwin's 'dilemma' 22, 25–6
David 47, 84–5, 90, 100
Dead Sea Scrolls 42, 61–2
Devil, the 88, 105
dinosaurs 19, 20, 29–8
disciples 80, 81, 97
DNA 13–14, 17–18, 19, 29, 47
    see also mutation

Earth 3
    creation of 1, 2, 11, 35–7
        see also Big Bang Theory; Creationism
    destruction of 6–7, 53
    gravitational pull 5
    meteorite collision 7, 28
Easter 70
Eber 58
Egypt, ancient 60, 68–70, 71, 75, 79
Einstein, Albert 9
Elijah 86

Elizabeth (mother of John the Baptist) 86
Enoch 57, 59
Enosh 57
Esau 58
Eve 37, 42, 43–4, 48–52, 59, 105
evil, force of 44, 104–5
Evolution, Theory of 12–13, 14, 18–21, 22
    see also Darwin, Charles
Exodus, Book of 59, 111

Faraday, Michael 9
Flood, the Great 30–31, 58, 59, 65–8

Galilee 71, 86, 98, 115
Galileo 9
Genesis, Book of 45, 57–9, 65, 88
    quotations from 41, 43, 49, 68, 100, 108, 121
Gentiles 80, 92–3, 99
God 108, 121–2
    creation of Adam and Eve 42–4, 49–50, 51
    creation of time 50–51
    and the Holy Spirit 113–15, 117–18
good, force of 105
grasses 17
gravity 2, 4–6
Great Flood, the 30–31, 58, 59, 65–8

Greece, ancient 69, 95, 96

heat 6–7, 12
heaven 97, 122
Herod Agrippa I, King of Judea 94
Herod Antipas, Ruler of Galilee & Perea 86, 98
Herodotus 79
Holy Spirit 45, 88, 99, 108, 110–18
Homo sapiens 30, 36
human beings 7
  see also Creationism; Evolution Theory of
  lifespans of 68
  'modern' (Homo sapiens) 30, 36, 40, 46, 54, 124
  primitive 21–2, 28–9, 40, 49, 52, 104
  survival of the Great Flood 65–6
human body 4, 6, 15–16, 107–8
  see also DNA
  the mind 106–7
  spirit 52–4, 108–10
  the soul 52–4, 108–10, 113, 116
Human Genome Project, The 8
Huxley, Thomas 22

ice ages 12, 20, 28, 29–30, 36, 66–7

Incas 78
Isaac 58, 74
Isaiah, Book of 62, 84, 85, 87, 89
Islam 25, 42, 83, 96, 111, 121
Israel 46, 64, 85, 92
Israelites 70, 71, 74–6, 84, 111

Jacob 59, 68, 74
Jared 57
Jashar, Book of 75
Jericho 74–75
Jerusalem 64, 90, 93, 95, 100, 115
Jesus Christ 42, 44, 45, 63, 80–89, 96–7
  and the Holy Spirit 108, 112, 117
  miracles with water 71
  Resurrection of 90–91, 99–103, 101
John the Apostle 80–81, 89, 96, 97, 112, 117
John the Baptist 86–7, 98, 112
Jordan, River 74–5
Joseph (father of Jesus) 85
Josephus 98–9
Joshua 74–5, 77, 85
Judaism 25, 28, 64, 70, 96, 98
  genealogy 40, 42, 46–7, 57, 83–4, 100, 121
  see also Torah, the
Judas 81

Kenan 57, 59
Kohath 59
Koran, the 83

Lamech 57, 59
language 41–2, 49, 115
Last Common Ancestor (LCA) 22
Lebanon 96
Levi 59
levitation 119
light 6–7
Luke 80–81, 85, 86, 91–2, 93, 95, 96, 97, 113

Mahalalel 57
Maiden Castle 31
Malta 91, 94, 95, 96
Mark 80–81, 86, 96, 97
marriage 46–7
Mary, mother of Jesus 47, 71, 84, 85
Matthew 80–81, 84, 96, 97
Maxwell, John Clerk 9
McDowell, Josh 62
Mesopotamia 42, 66, 115
Methuselah 57, 58, 59
Meyer, Stephen C, *Darwin's Doubt* 26
Michelangelo 111
mind, the 106–7
Minoan eruption 69–70
miracles 71–4, 78, 97, 110, 119
missing day 77–9

'missing link' 22, 45
Moab, ancient 60
moon 4, 5–6, 75
Morison, Frank, *Who Moved the Stone* 100–101
Moses 46, 47, 49, 56, 57, 59–60, 70, 74, 85, 111
mutation 14, 19, 20, 23
  see also DNA

Nahar 58
Natural Selection 17, 18–21, 23, 24, 108
Nebo, Mount 74
Nero, Emperor of Rome 95
Newton, Isaac 9
Nile, River 69, 71
Noah 30, 46, 47, 56–7, 58, 59, 65–6, 110
  see also Great Flood, the

oceans 12
Oholiab 111
out-of-body experiences 122
ozone layer 6

parables 97
Paul the Apostle 81, 91–6
Peleg 58
Pentateuch 28, 49
  see also Exodus, Book of; Genesis, Book of
Persia, ancient 78
Peter 105
Philips, William D 8

Pilate, Pontius 88, 99
plagues 68–70
Polkinghorne, Sir John 8
prayer 53, 108, 116
primates *see* apes; chimpanzees
prophecies 63–4, 83–5, 87, 89–90, 96, 98, 100, 111
Psalms, Book of 89, 90

rainbows 4, 67
Red Sea 70, 75
Reu 58
Roman Empire 82, 86, 95, 100
   Pontius Pilate 88, 99
   and Paul the Apostle 91, 93, 94–5
Royal Society 8

Santorini, Greece 69–70
Saul (Paul the Apostle) 81, 91–6
Serug 58
Seth 57
Shem 58
Shlar 58
Skara Brea 31
Solar System 2, 3, 6, 13, 24
soul, the 52–4, 108–10, 113, 116
speaking in tongues 115–17
spiritual life 121–2

Stonehenge 31
sun 4, 5–6, 75
supernatural events 65, 119–20
Syria 96

Tacitus 95
Terah 58
time 1–2, 41, 50–51, 77–9, 124–5
Titan 24
tongues, speaking in 115–17
Torah, the 84, 85, 86, 88
   Pentateuch 28, 49
      see also Genesis, Book of; Exodus
Turkey 93, 96

Universe 1–2, 11, 13

water 3, 4, 12, 71–4
   see also Great Flood, the

Wilson, Robert Dick, *A Scientific Investigation of the Old Testament* 60–61, 62
written records 41–2, 56–7, 60

Zechariah (father of John the Baptist) 86
Zechariah, Book of 64, 89, 90

Lightning Source UK Ltd.
Milton Keynes UK
UKHW010237291020
372433UK00001B/137